Computational Approaches in Drug Discovery and Design

Computational Approaches in Drug Discovery and Design

Guest Editor

Shijun Zhong

MDPI

Basel • Beijing • Wuhan • Barcelona • Belgrade • Novi Sad • Cluj • Manchester

Guest Editor
Shijun Zhong
School of Bioengineering
Dalian University of Technology
Dalian
China

Editorial Office
MDPI AG
Grosspeteranlage 5
4052 Basel, Switzerland

This is a reprint of the Special Issue, published open access by the journal *Molecules* (ISSN 1420-3049), freely accessible at: https://www.mdpi.com/journal/molecules/special_issues/72RI9ZRI13.

For citation purposes, cite each article independently as indicated on the article page online and as indicated below:

Lastname, A.A.; Lastname, B.B. Article Title. *Journal Name* **Year**, *Volume Number*, Page Range.

ISBN 978-3-7258-6702-8 (Hbk)
ISBN 978-3-7258-6703-5 (PDF)
https://doi.org/10.3390/books978-3-7258-6703-5

Contents

About the Editor

Shijun Zhong

Shijun Zhong is a professor of computational chemistry at Dalian University of Technology, China. He received his PhD degree in 1994 from Xiamen University and then worked as a postdoctoral fellow at the Fujian Institute of Research on the Structures of Matter, the Chinese Academy of Sciences. In 1996, he became an associate professor at Xiamen University. Before moving to Dalian University of Technology in 2011, he spent thirteen years performing research in some overseas institutes, including the Hebrew University of Jerusalem, the Lawrence Berkeley National Laboratory, Wesleyan University, and the University of Maryland, Baltimore. His research interests include some subfields of quantum chemistry, molecular dynamics simulation, and virtual molecular database screening, such as the irreducible tensor sets for the construction of wavefunctions, symmetry simplification of the two-electron integral calculations, fragmentation of protein electrostatic potentials, Gaussian basis sets, molecular force field parameters for drug molecules, prediction of ligand binding sites, and scaling of binding poses for drug design. He performed computations on a number of molecular systems, including the FeMo-cofactor of nitrogenase, heteroatom fullerene cages, transition-metal clusters, oligosaccharides, peptides, and proteins. He also designed inhibitors and binders for human DNA ligase 1, leucine-rich repeat kinase 2, tyrosine phosphatase 1B, β-lactamase, flocculation protein Flo1p, and Ubiquitin-Specific Protease-7.

molecules

MDPI

Article

Pairwise Performance Comparison of Docking Scoring Functions: Computational Approach Using InterCriteria Analysis

Maria Angelova, Petko Alov, Ivanka Tsakovska, Dessislava Jereva, Iglika Lessigiarska, Krassimir Atanassov, Ilza Pajeva and Tania Pencheva *

Institute of Biophysics and Biomedical Engineering, Bulgarian Academy of Sciences, 105 Acad. G. Bonchev Str., 1113 Sofia, Bulgaria; maria.angelova@biomed.bas.bg (M.A.); petko@biophys.bas.bg (P.A.); itsakovska@biomed.bas.bg (I.T.); dessislava.jereva@biomed.bas.bg (D.J.); iglika@biomed.bas.bg (I.L.); k.t.atanassov@gmail.com (K.A.); pajeva@biomed.bas.bg (I.P.)

* Correspondence: tania.pencheva@biomed.bas.bg

Abstract

Scoring functions are key elements in docking protocols as they approximate the binding affinity of a ligand (usually a small bioactive molecule) by calculating its interaction energy with a biomacromolecule (usually a protein). In this study, we present a pairwise comparison of scoring functions applying a multi-criterion decision-making approach based on InterCriteria analysis (ICrA). As criteria, the five scoring functions implemented in MOE (Molecular Operating Environment) software were selected, and their performance on a set of protein–ligand complexes from the PDBbind database was compared. The following docking outputs were used: the best docking score, the lowest root mean square deviation (RMSD) between the predicted poses and the co-crystallized ligand, the RMSD between the best docking score pose and the co-crystallized ligand, and the docking score of the pose with the lowest RMSD to the co-crystallized ligand. The impact of ICrA thresholds on the relations between the scoring functions was investigated. A correlation analysis was also performed and juxtaposed with the ICrA. Our results reveal the lowest RMSD as the best-performing docking output and two scoring functions (Alpha HB and London dG) as having the highest comparability. The proposed approach can be applied to any other scoring functions and protein–ligand complexes of interest.

Keywords: molecular operating environment; scoring functions; molecular docking; InterCriteria analysis

1. Introduction

The search for hit compounds with specific biological activity requires considerable effort and resources. The primary efforts concern the synthesis and biological testing of a large number of compounds. This defines the strong interest of the R&D pharmaceutical sector towards computer-aided methods that facilitate rational drug design [1]. Among these methods, molecular docking is particularly important when the structure of a biomacromolecule is known. This method is used to predict the binding affinity of a ligand in the active site (by employing scoring functions to calculate docking scores) based on the ligand docking poses (conformations and positions) in the binding site. Docking is also applied in virtual screening to identify potential active ligands for a specific macromolecular target among thousands to millions of compounds. As the calculations of docking scores can be time-consuming, the selection of appropriate scoring functions could impact the overall effectiveness of the computational procedure. Over the past decades, a number of studies have conducted comparative assessments of scoring functions implemented in both free and commercial docking software [2–5]. To evaluate the predictive power and examine the strengths and limitations of the scoring functions, several benchmark test sets have been developed [3,6,7]. These datasets are highly diverse, encompassing a wide range of protein families, ligand chemotypes, and binding affinities [8], thus providing good opportunities for the comparison of docking scoring functions. Despite the studies [2–5] assessing and comparing the scoring functions in docking, the question regarding which functions are more appropriate for use compared to others remains open.

In this study, the docking scoring functions implemented in Molecular Operating Environment (MOE, https://www.chemcomp.com/, accessed on 22 May 2025) [9], the widely used drug discovery software platform, were selected for comparative analysis. The MOE docking module operates with five scoring functions, namely London dG, ASE, Affinity dG, Alpha HB, and GBVI/WSA dG [8,9]. The first four are empirical, while GBVI/WSA dG is a force-field function [5]. Empirical functions evaluate the binding affinity of protein–ligand complexes based on a set of weights defined by multiple linear regression to experimentally measured affinities. The equation terms describe important contributions to the protein–ligand binding, like hydrogen bonding, ionic and hydrophobic interactions, loss in ligand flexibility, etc. On the other hand, the force-field-based scoring functions use classical force fields to evaluate protein–ligand interactions. In their simplest forms, Lennard-Jones and Coulomb potentials are applied to describe the enthalpy terms.

The comparison of MOE scoring functions in this research is based on the Comparative Assessment of Scoring Functions benchmark subset (CASF-2013) [3] of the PDBbind database [6], which is a comprehensive collection of 195 protein–ligand complexes with available binding affinity data. The CASF-2013 dataset was used for the assessment of scoring functions implemented in free docking software such as AutoDock and Vina [10]. In [11], the CASF-2013 dataset supported the evaluation of 12 scoring functions based on machine learning techniques. The high-quality set of 195 test complexes enabled the comparison and assessment of scoring functions embedded in leading software platforms like Schroedinger, MOE, Discovery Studio, SYBYL, and GOLD [12]. CASF-2013, along with another independent benchmark dataset named Community Structure-Activity Resource (CSAR 2014), was used in an evaluation analysis of an improved version of the scoring function SPecificity and Affinity (SPA) [7]. CASF-2013 was applied to validate 3D convolutional neural network (3D-CNN) scoring functions and end-to-end 3D-CNNs with a mechanism for spatial attention [13,14], as well as for extensive validation of the scoring power of geometric graph learning with extended atom-type features for protein–ligand binding affinity prediction [15]. Thus, CASF-2013 was selected as the dataset in our study.

The InterCriteria analysis (ICrA) [16] was applied for pairwise performance comparison of the MOE scoring functions on CASF-2013. The expectations were for the ICrA to reveal relations between the scoring functions by analyzing a variety of data outputs extracted from molecular docking. ICrA was elaborated as a multi-criterion decision-making approach that detects possible relations between pairs of criteria when multiple objects are considered. Over the past years of research, the approach has already demonstrated its potential when applied to economic, biomedical, industrial, and various other data sets and problem formulations [17]. One of the research objectives was to specify the general framework of the problems that may be solved by the approach, as well as the connections between the new approach and classical multi-criteria decision-making methods such as correlation analyses [https://intercriteria.net, accessed on 22 May 2025]. Some case studies of the ICrA applications are freely accessible [https://intercriteria.net/studies/, accessed on 22 May 2025], e.g., in relation to the World Economic Forum's global competitiveness reports, genetic algorithm performance, ant colony optimization performance, and analysis of calorimetric data of blood serum proteome.

Recently, the ICrA approach was applied in the field of computer-aided drug design and computational toxicology in comparative studies of various scoring functions [18,19], as well as in in silico studies of biologically active molecules [20]. In contrast to our previous studies, here we apply ICrA on the CASF-2013 benchmark subset of the PDBbind database covering several important docking outputs. The current investigation is a significant extension of our previous studies [18–20] as it is based on a much larger dataset of structurally and functionally different proteins and their ligands, with a detailed investigation of the impact of key ICrA thresholds on the comparison results. A correlation analysis was also performed and juxtaposed with the ICrA results. Our results outline the most and the least similar scoring functions and the most comparable docking outputs. In addition, they demonstrate the applicability of ICrA to reveal new relations between the studied criteria.

2. Results and Discussion

For the CASF-2013 dataset, re-docking of the ligands in the protein–ligand complexes was performed. The following data were extracted from 30 saved poses:

(i) the best docking score (a lower score suggests better protein–ligand binding)—hereinafter referred to as BestDS;
(ii) the lowest root mean square deviation (RMSD) between the predicted poses and the ligand in the co-crystallized complex—hereinafter referred to as BestRMSD;
(iii) the RMSD between the pose with the best docking score and the ligand in the co-crystallized complex—hereinafter referred to as RMSD_BestDS;
(iv) the docking score of the pose with the lowest RMSD to the ligand in the co-crystallized complex—hereinafter referred to as DS_BestRMSD.

For the sake of completeness, we have also investigated the possible relations to the experimental data available in CASF-2013. Based on the availability, K_d or K_i, expressed as $(-\log K_d)$ or $(-\log K_i)$, respectively, were used.

The collected information was formatted for subsequent application of the ICrA: the ligands from protein–ligand complexes were considered objects of research in terms of the ICrA, while the different scoring functions (represented by BestDS, BestRMSD, RMSD_BestDS, or DS_BestRMSD), along with the binding affinity data, were considered as criteria.

ICrA was applied in two steps. Initially, it was implemented on the data outputs extracted from molecular docking results as described above, with the conditionally defined (default in ICrAData) values $\alpha = 0.75$ and $\beta = 0.25$, based on which the ranges of *consonance* and *dissonance* are defined [21].

Table 1 summarizes the results obtained for the degrees of agreement μ between the five scoring functions of MOE and the binding affinity data of the studied complexes after the ICrA application. The colors in Table 1 reproduce the ones used in ICrAData (green represents *positive consonance*, magenta represents *dissonance*) for the conditionally defined values of $\alpha = 0.75$ and $\beta = 0.25$. The ICrA relations between the scoring functions and experimental data are highlighted in grey at the bottom of the table.

As seen from Table 1, "varicolored" ICrA results were obtained only when BestRMSD was considered. For the rest of the docking outputs, ICrA does not outline any significant relations (other than in *dissonance*) between the five investigated scoring functions in MOE, as well as between the scoring functions and the binding affinity data—all hit the *dissonance* zone only (colored in magenta).

Table 1. Degrees of agreement μ obtained after ICrA application with the conditionally defined ICrAData threshold values of α = 0.75 and β = 0.25.

	BestDS	BestRMSD	RMSD_BestDS	DS_BestRMSD
Affinity dG–Alpha HB	0.60	0.81	0.67	0.59
Affinity dG–ASE	0.62	0.77	0.68	0.57
Affinity dG–GBVI/WSA dG	0.55	0.83	0.67	0.61
Affinity dG–London dG	0.56	0.78	0.63	0.56
Alpha HB–ASE	0.66	0.79	0.64	0.62
Alpha HB–GBVI/WSA dG	0.47	0.76	0.69	0.45
Alpha HB–London dG	0.72	0.84	0.68	0.70
ASE–GBVI/WSA dG	0.44	0.73	0.66	0.36
ASE–London dG	0.62	0.77	0.65	0.60
GBVI/WSA dG–London dG	0.48	0.73	0.64	0.46
Affinity dG–(−logK_d) or (−logK_i)	0.45	0.57	0.50	0.53
Alpha HB–(−logK_d) or (−logK_i)	0.40	0.53	0.44	0.49
ASE–(−logK_d) or (−logK_i)	0.35	0.56	0.37	0.57
GBVI/WSA dG–(−logK_d) or (−logK_i)	0.58	0.56	0.60	0.55
London dG–(−logK_d) or (−logK_i)	0.45	0.53	0.45	0.48

In the second step, the impact of the variations in α and β values on the investigated relations was explored. The thresholds α and β might be determined algorithmically or chosen by the user. Since these parameters define the thresholds for *positive consonance/dissonance/negative consonance* between the studied pairs of criteria, we investigated the impact of their values on the relations between the scoring functions. We varied the difference between α and β (keeping their sum at 1.00) starting from 0.5 (α = 0.75 and β = 0.25) and decreasing it gradually to 0.2 (α = 0.6 and β = 0.4) to follow how these values affect the relations between the investigated scoring functions (for clarity, these analyses are reported using ICrA colors in Supplementary Tables S1–S5). The results are summarized in Tables 2 and 3 based on the following two indicators: the number of pairs of scoring functions in *positive consonance* for each docking output (Table 2) and the number of docking outputs in *positive consonance* for each pair of scoring functions (Table 3).

Table 2. Number of scoring function pairs in *positive consonance* for each docking output.

α/β	BestDS	BestRMSD	RMSD_BestDS	DS_BestRMSD
0.75/0.25	0	8	0	0
0.70/0.30	1	10	0	1
0.67/0.33	1	10	5	1
0.65/0.35	2	10	7	1
0.60/0.40	5	10	11	4

The first indicator, the number of pairs of scoring functions in *positive consonance*, could be considered as a measure of the "sensitivity" of the corresponding docking output to the α and β values. The second one, the number of docking outputs for each pair in *positive consonance*, allows for a comparison of the scoring functions based on the number of the docking outputs—a higher number of docking outputs in *positive consonance* indicates a more similar performance of the pairs of scoring functions studied.

As expected, the decrease in the difference between α and β results in a higher number of pairs in *positive consonance*, thus raising the question regarding the most appropriate α and β values. The values of 0.67/0.33 and 0.65/0.35 show almost identical results, and 0.67/0.33 appears to be a good compromise between "no" and "many" pairs in *positive consonance*, also allowing for a certain tolerance in comparability between the studied criteria.

Table 3. Number of docking outputs in *positive consonance* for each pair of scoring functions.

Pairs of Scoring Functions	α/β				
	0.75/0.25	0.70/0.30	0.67/0.33	0.65/0.35	0.60/0.40
Affinity dG–Alpha HB	1	1	2	2	3
Affinity dG–ASE	1	1	2	2	3
Affinity dG–GBVI/WSA dG	1	1	2	2	3
Affinity dG–London dG	1	1	1	1	2
Alpha HB–ASE	1	1	1	2	4
Alpha HB–GBVI/WSA dG	1	1	2	2	2
Alpha HB–London dG	**1**	**3**	**4**	**4**	**4**
ASE–GBVI/WSA dG		1	1	2	2
ASE–London dG	1	1	1	2	4
GBVI/WSA dG–London dG		1	1	1	2
Affinity dG–($-\log K_d$) or ($-\log K_i$)	0	0	0	0	0
Alpha HB–($-\log K_d$) or ($-\log K_i$)	0	0	0	0	0
ASE–($-\log K_d$) or ($-\log K_i$)	0	0	0	0	0
GBVI/WSA dG–($-\log K_d$) or ($-\log K_i$)	0	0	0	0	1
London dG–($-\log K_d$) or ($-\log K_i$)	0	0	0	0	0

In addition, the values $\alpha = 0.67$ and $\beta = 0.33$ allow for the presentation of an alternate scale for *consonance/dissonance*. If the scale of *consonance/dissonance* outlined in Atanassov et al. [21] might be considered as a scale of "quarters", here a kind of scale of "thirds" is intentionally considered, corresponding to $\alpha = 0.67$ (approximately two-thirds) and $\beta = 0.33$ (approximately one-third). For better understanding, Figure 1 represents the respective interpretative intuitionistic triangle.

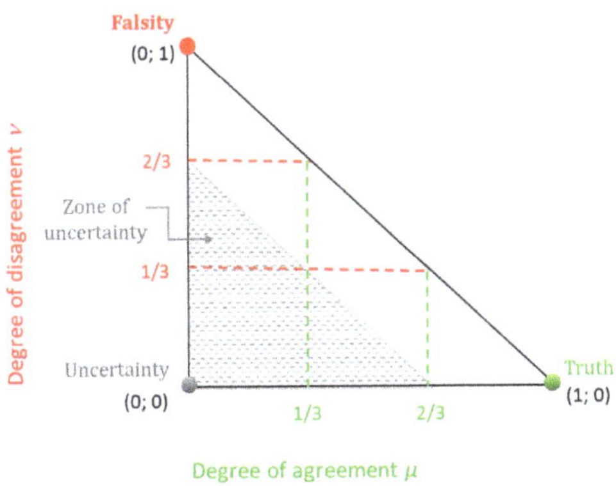

Figure 1. Interpretative intuitionistic triangle in the case of "thirds".

In this case, the pair of criteria are said to be in one of the following:

- *positive consonance*, whenever $\mu_{C_k,C_l} \geq 2/3$ and $\nu_{C_k,C_l} < 1/3$;
- *negative consonance*, whenever $\mu_{C_k,C_l} < 1/3$ and $\nu_{C_k,C_l} \geq 2/3$;
- *dissonance*, whenever $0 \leq \mu_{C_k,C_l} < 2/3$, $0 \leq \nu_{C_k,C_l} < 2/3$, and $2/3 \leq \mu_{C_k,C_l} + \nu_{C_k,C_l} \leq 1$; and
- *uncertainty*, $0 \leq \mu_{C_k,C_l} < 2/3$, $0 \leq \nu_{C_k,C_l} < 2/3$, and $0 \leq \mu_{C_k,C_l} + \nu_{C_k,C_l} < 2/3$.

As seen in Figure 1, the presented scale of "thirds" preserves the symmetry in the interpretative intuitionistic triangle in accordance with the original "quarters"-based symmetry [21]. Thus, the combination $\alpha = 0.67$ and $\beta = 0.33$ appears to be an appropriate selection in this investigation. It should be noted that such a systematic investigation of the impact of different threshold values for the *consonance* and *dissonance* intervals in the ICrA is conducted for the first time in the current study and may be considered as a methodological contribution to the presented analysis.

According to the pairwise performance comparison of the MOE scoring functions (Table 3), the following results can be outlined:

(1) an absence of any kind of agreement in all explored values of α and β with the experimental data (except one, between GBVI/WSA dG and ($-\log K_d$) or ($-\log K_i$) at $\alpha = 0.60$ and $\beta = 0.40$, rows highlighted in grey in Table 3). This result is in accordance with our previous studies [19]. The lack of any agreement might also be explained by the fact that, even when implementing a variety of scoring terms and becoming more sophisticated, the scoring functions are still a computational approximation mostly aimed at assisting in the prediction of ligand binding poses. This is confirmed by the results of the BestRMSD docking output (Table 2).

(2) a *positive consonance* between two scoring functions, Alpha HB and London dG: in particular, for 0.67/0.33 threshold values, they are comparable in all four docking outputs (a row in bold in Table 3). The result suggests that these scoring functions might be used interchangeably. At the same time, some pairs show small comparability (Affinity dG–London dG and GBVI/WSA dG–London dG), suggesting that they can complement each other in consensus docking studies.

Figures 2 and 3 demonstrate the results from ICrA implementation for the explored values of thresholds α and β, respectively, for BestDS and RMSD_BestDS. Both figures show the ICrAData screenshots at conditionally defined threshold values of $\alpha = 0.75$ and $\beta = 0.25$ (subplot a), $\alpha = 0.70$ and $\beta = 0.30$ (subplot b), $\alpha = 0.67$ and $\beta = 0.33$ (subplot c), $\alpha = 0.65$ and $\beta = 0.35$ (subplot d), and $\alpha = 0.60$ and $\beta = 0.40$ (subplot e). Subplot (c) represents the full screenshot from the ICrAData software, while the other subplots represent only the results for the degrees of agreement μ and the intuitionistic fuzzy triangle at different values of thresholds.

As seen in Figure 2, applying new values of thresholds leads to the appearance of pairs in *positive consonance*. For $\alpha = 0.70$ and $\beta = 0.30$ (Figure 2b) and $\alpha = 0.67$ and $\beta = 0.33$ (Figure 2c), *positive consonance* appears between the scoring functions Alpha HB and London dG, with no identified significant relations with the conditionally defined threshold values (Figure 2a). Further decreasing the difference between α and β leads to one more pair in *positive consonance*—Alpha HB–ASE for $\alpha = 0.65$ and $\beta = 0.35$ (Figure 2d), and additionally three more pairs in *positive consonance*—namely Affinity dG–Alpha HB, Affinity dG–ASE, and ASE–London dG at $\alpha = 0.60$ and $\beta = 0.40$ (Figure 2e).

As seen in Figure 3 and Table 1, RMSD_BestDS has the highest number of significant relations that appear when the new values of the thresholds are applied. Altogether, five pairs of scoring functions show *positive consonance* at $\alpha = 0.67$ and $\beta = 0.33$, namely, Affinity dG–Alpha HB, Affinity dG–ASE, Affinity dG–GBVI/WSA dG, Alpha HB–GBVI/WSA dG, and Alpha HB–London dG (Figure 3c), in comparison to no significant relations identified at the conditionally defined threshold values (Figure 3a) of $\alpha = 0.70$ and $\beta = 0.30$ (Figure 3b). Further decreasing the difference between α and β leads to two more pairs in *positive consonance*—between ASE–GBVI/WSA dG and ASE–London dG at $\alpha = 0.65$ and $\beta = 0.35$ (Figure 3d), and even to three additional pairs between Affinity dG–London dG, Alpha HB–ASE, and GBVI/WSA dG–London dG at $\alpha = 0.60$ and $\beta = 0.40$ (Figure 3e).

As mentioned above, the most "varicolored" picture from ICrA implementation is when the BestRMSD is considered (the results are presented in Supplementary Tables S1–S5). As seen in Table 1, *positive consonance* was observed for almost all pairs of scoring functions, while the other two pairs of scoring functions, ASE–GBVI/WSA dG and

GBVI/WSA dG–London dG, are very close to the range of *positive consonance*. Then, still at the threshold values α = 0.70 and β = 0.30, all pairs of scoring functions show *positive consonance*. Based on this analysis, one may conclude that, according to the BestRMSD, all scoring functions give quite similar results.

For the DS_BestRMSD output, only one pair of scoring functions, Alpha HB–London dG, falls into the interval of *positive consonance* (Tables 1 and 2; whole information is available in Supplementary Tables S1–S5,) when applying values of thresholds α = 0.70 and β = 0.30. Further decreases in the difference between α and β lead to three more pairs in *positive consonance*—between Affinity dG–GBVI/WSA dG, Alpha HB–ASE, and ASE–London dG, only at α = 0.60 and β = 0.40.

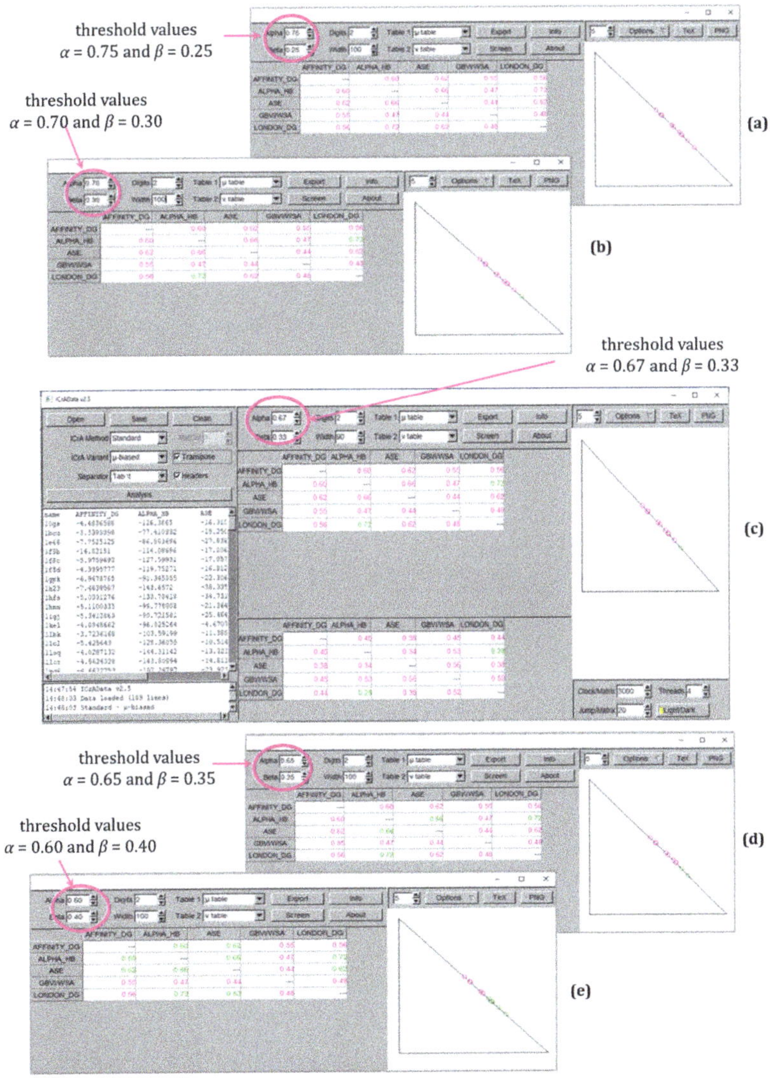

Figure 2. ICrA implementation with different values of α and β thresholds to assess the scoring functions based on BestDS as follows: (**a**) α = 0.75 and β = 0.25; (**b**) α = 0.70 and β = 0.30; (**c**) α = 0.67 and β = 0.33, (**d**) α = 0.65 and β = 0.35; and (**e**) α = 0.60 and β = 0.40.

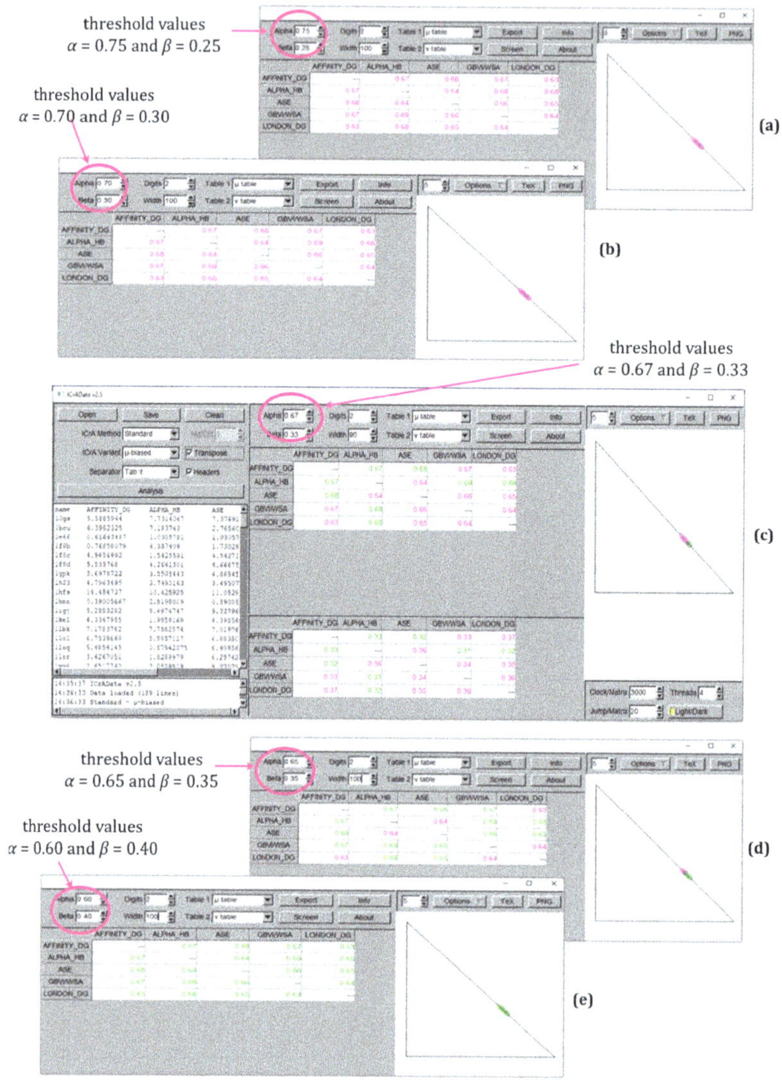

Figure 3. ICrA implementation with different values of α and β thresholds to assess the five scoring functions based on RMSD_BestDS as follows: (**a**) $\alpha = 0.75$ and $\beta = 0.25$; (**b**) $\alpha = 0.70$ and $\beta = 0.30$; (**c**) $\alpha = 0.67$ and $\beta = 0.33$, (**d**) $\alpha = 0.65$ and $\beta = 0.35$; and (**e**) $\alpha = 0.60$ and $\beta = 0.40$.

Positive consonance between the scoring functions Alpha HB and London dG was outlined in all investigated docking outputs. The close comparability between London dG and Alpha HB could be related to the fact that both functions include explicit terms related to hydrogen bonding. This type of protein–ligand interaction is among the most common, and it typically has a significant contribution to the binding energy of the protein–ligand complex. In addition, both scoring functions include terms that implicitly account for the conformational changes in the ligands upon binding to achieve the best fit of the ligand in the protein active site. In the case of London dG, this is estimated by the loss of ligand flexibility (calculated from ligand topology only); for Alpha HB, the geometrical fit to the protein active site is estimated by summing up the attractive and repulsive interactions between the protein and ligand atoms. In addition, London dG accounts for metal ligation and certain entropic effects (the average gain/loss of rotational and translational entropy and desolvation effects). The decision on which scoring function to use could be made based on the preliminary analysis of the protein–ligand interactions in the experimental complex of interest.

For the completeness of the comparison of the five scoring functions, a correlation analysis (CA) has also been performed. Table 4 summarizes the results obtained by the ICrA (the degrees of agreement μ are reported) and CA (Pearson correlation coefficient R) for all docking outputs. The CA shows a higher correlation only for BestRMSD, while for the other docking outputs, the observed correlations are relatively low. In the case of BestRMSD, the highest correlation coefficients coincide with the highest values of ICrA degrees of agreement. In particular, the relation between Alpha HB and London dG is evaluated with the highest values of degree of agreement by ICrA and with the second-best correlation coefficient by CA. Figure 4 illustrates the correlation in terms of CA between Alpha HB and London dG for BestRMSD (Table 4, correlation coefficient 0.79).

Table 4. Results obtained after ICrA and CA applications.

	BestDS		**BestRMSD**		**RMSD_BestDS**		**DS_BestRMSD**	
	ICrA μ	CA R	ICrA μ	CA R	ICrA μ	CA R	ICrA μ	CA R
Affinity dG–Alpha HB	0.60	0.20	0.81	0.74	0.67	0.55	0.59	0.19
Affinity dG–ASE	0.62	0.23	0.77	0.68	0.68	0.53	0.57	0.03
Affinity dG–GBVI/WSA dG	0.55	0.12	0.83	0.81	0.67	0.55	0.61	0.22
Affinity dG–London dG	0.56	0.14	0.78	0.66	0.63	0.35	0.56	0.06
Alpha HB–ASE	0.66	0.54	0.79	0.77	0.64	0.45	0.62	0.42
Alpha HB–GBVI/WSA dG	0.47	−0.10	0.76	0.63	0.69	0.53	0.45	−0.05
Alpha HB–London dG	0.72	0.55	0.84	0.79	0.68	0.52	0.70	0.47
ASE–GBVI/WSA dG	0.44	−0.19	0.73	0.57	0.66	0.48	0.36	−0.16
ASE–London dG	0.62	0.29	0.77	0.68	0.65	0.42	0.60	0.24
GBVI/WSA dG–London dG	0.48	−0.16	0.73	0.57	0.64	0.36	0.46	−0.06
Affinity dG–(−logK_d) or (−logK_i)	0.45	−0.08	0.57	0.21	0.50	0.19	0.53	0.06
Alpha HB–(−logK_d) or (−logK_i)	0.40	−0.29	0.53	0.10	0.44	0.03	0.49	−0.18
ASE–(−logK_d) or (−logK_i)	0.35	−0.42	0.56	0.12	0.37	0.24	0.57	−0.37
GBVI/WSA dG–(−logK_d) or (−logK_i)	0.58	0.13	0.56	0.23	0.60	0.19	0.55	0.09
London dG–(−logK_d) or (−logK_i)	0.45	−0.13	0.53	0.06	0.45	−0.02	0.48	−0.11

The absence of correlation (estimated by the Pearson correlation coefficients) between the docking scores and the experimental data on the binding affinity of the ligands in the complexes is not surprising. This result could not be related only to the heterogeneity of the experimental data (different measures of affinity, different methods and experimental protocols, various protein–ligand complexes) in the dataset used but rather to the fact that molecular docking has not initially been designed for the correlation of docking scores with experimental binding affinities [22]. The absence of such correlations has also been confirmed in our previous studies employing more consistent experimental data on the binding affinities of a homologous series of 88 benzamidine-type ligands toward thrombin, trypsin, and factor Xa [19]. Similar findings have been reported in systematic evaluations of scoring functions by other research groups. Notably, Li et al. [12] assessed a panel of 20 scoring functions, including those implemented in MOE, using the same CASF-2013 dataset. They concluded that the investigated scoring functions generally performed better at predicting binding poses than affinity assessments. These results are in accordance with our findings on the best-performing docking output (BestRMSD) and the absence of consonance with the experimental values; however, in contrast to [12], we confirmed these results by estimating similarity/dissimilarity between the scoring functions. In this way, our analysis adds a new value to the comparison of the scoring functions.

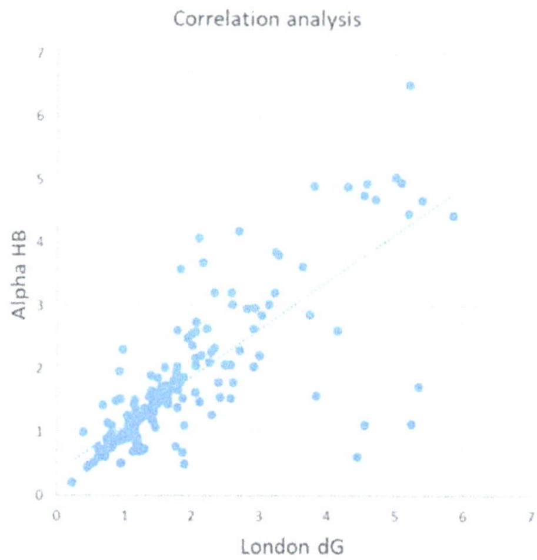

Figure 4. Correlation plot for Alpha HB and London dG based on the BestRMSD.

As seen in Table 4 and Figure 4, ICrA may offer enhanced capabilities over CA by enabling the identification of additional relationships among the evaluated scoring functions. ICrA, as well as CA, reports the degree of coincidence (in terms of ICrA, this is *positive consonance*), and both CA and ICrA report negative correlation (*negative consonance* in terms of ICrA), in which the values for one criterion increase while at the same time the values of the other criterion decrease. Unlike CA, ICrA also allows for classifying the criteria relations in *dissonance*, as well as accounting for the degree of uncertainty, which represents the advantage of ICrA over CA.

In this way, the results from ICrA implementation for the pairwise performance comparison of docking scoring functions would allow the user to decide on the selection of an appropriate scoring function in order to optimize the computational costs. This is relevant, especially when applied to virtual screening tasks. If two or more scoring functions produce similar results, only one of them can be used; instead, the scoring functions that give different results can be combined in consensus docking studies.

3. Materials and Methods

3.1. Dataset

In the current investigation, CASF-2013 was selected as a benchmark dataset widely recognized by researchers dealing with the development and assessment of scoring functions in structure-based studies due to its representativeness, diversity, and quality of complex structures and binding affinity data [3] (Figure 5).

The final CASF-2013 dataset consists of 195 protein–ligand complexes with binding affinity data selected out of 8302 protein–ligand complexes recorded in the PDBbind database (https://www.pdbbind-plus.org.cn/casf, v. 2013, accessed on 22 May 2025). The qualified complexes were classified in 65 clusters by 90% similarity in protein sequences. Three representative complexes are chosen from each cluster to control sample redundancy.

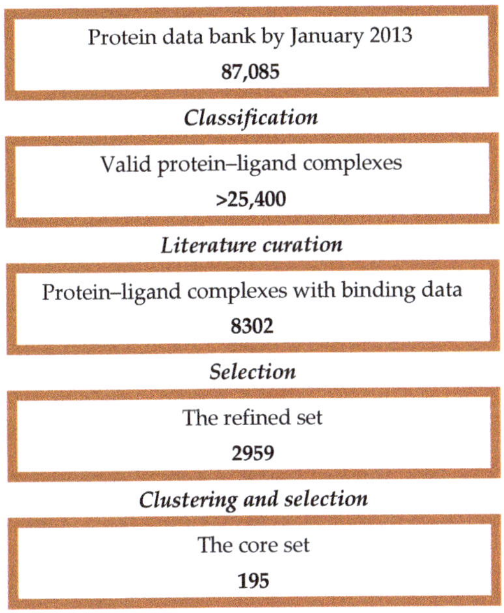

Figure 5. CASF-2013 dataset selection (adapted from [3]).

3.2. Molecular Docking

MOE v. 2022 was used for the molecular docking studies. Prior to re-docking, all 195 protein–ligand complexes were protonated using the Protonate3D tool in MOE. The tool assigns hydrogens to structures following the proton geometry with optimal free energy and the ionization states of titratable protein groups. The rigid protein/flexible ligand option was used. The active site was defined by ligand atoms, and the water molecules were removed from the binding sites of the studied proteins. The placement was obtained by a triangle matcher algorithm.

Re-docking of the ligands in the protein–ligand complexes was performed, applying all available scoring functions in MOE v. 2022, namely London dG, ASE, Affinity dG, Alpha HB, and GBVI/WSA dG, as briefly described below [8,9]:

- ASE is based on the Gaussian approximation and depends on the radii of the atoms and the distance between the ligand atom and receptor atom pairs. ASE is proportional to the sum of the Gaussians over all ligand atom–receptor atom pairs.
- Affinity dG is a linear function that calculates the enthalpy contribution to the binding free energy, including terms based on interactions between H-bond donor and acceptor pairs, ionic interactions, metal ligation, hydrophobic interactions, unfavorable interactions (between hydrophobic and polar atoms,) and favorable interactions (between any two atoms).
- Alpha HB is a linear combination of two terms: (i) the geometric fit of the ligand to the binding site with regard to the attraction and repulsion depending on the distance between the atoms and (ii) H-bonding effects.
- London dG estimates the free binding energy of the ligand, accounting for the average gain or loss of rotational and translational entropy, the loss of flexibility of the ligand,

the geometric imperfections of H-bonds and metal ligations compared to the ideal ones, and the desolvation energy of atoms.

- GBVI/WSA dG estimates the free energy of ligand bindings considering the weighted terms for the Coulomb energy, solvation energy, and van der Waals contributions.

Up to 30 poses per ligand were saved for each of the protein–ligand complexes and each of the investigated scoring functions.

3.3. InterCriteria Analysis Approach

The ICrA approach developed by Atanassov et al. in 2014 [16] is based on two mathematical formalisms: index matrices (IMs) [23] and intuitionistic fuzzy sets (IFSs) [24]. The algebraic apparatus of IMs allows for the processing of data arrays of diverse dimensions, while the IFS is a mathematical tool for handling uncertainty. By relying on IMs and IFSs, the ICrA allows for the identification of intercriteria relations in terms of *consonance* or *dissonance* between each pair of criteria, thus differentiating from the classical correlation analysis.

In the concept of IFSs [17], Atanassov builds the ICrA on Zadeh's theory of fuzzy sets [25], which, on their side, are an extension of the classical notion of set. In classical set theory, an element either belongs or does not belong to the set; therefore, the membership of the element to the set is represented by the values 0 for non-membership or 1 otherwise. The fuzzy set theory introduces a degree of membership μ of the element x to the set, such that $\mu \in [0; 1]$. The theory of IFSs further expands this notion by including the degree of non-membership ν of the element x to the set, such that $\nu \in [0; 1]$.

In mathematical terms, set A is defined as an intuitionistic fuzzy set as follows:

$$A = \{\langle \mathrm{x}, \mu_{\mathrm{A}}(\mathrm{x}), \nu_{\mathrm{A}}(\mathrm{x}) \rangle \mid x \in X\},$$

where X is the universum, and the two mappings $\mu_{\mathrm{A}}(\mathrm{x})$, $\nu_{\mathrm{A}}(\mathrm{x})$: $A \to [0, 1]$ are, respectively, the degree of membership and the degree of non-membership of each element $x \in X$, such that $0 \leq \mu_{\mathrm{A}}(\mathrm{x}) + \nu_{\mathrm{A}}(\mathrm{x}) \leq 1$. The boundary conditions are 0 and 1, both giving the classical set.

The ICrA approach uses a two-dimensional (2D) IM as the input. The 2D IM is represented by a set indexing the rows, a set indexing the columns, and a set of elements corresponding to each pair of row and column index. In the case of ICrA, the IM can be written as $[O, C, e_{o,c}]$, where O and C are, respectively, the set of row and column indices, O stands for objects, C stands for criteria, and $e_{o,c}$ corresponds to the evaluation of each object O against the criterion C. If we denote m as the number of objects and n as the number of criteria, the input IM for the ICrA can be represented as follows:

	C_1	...	C_k	...	C_n
O_1	e_{O_1,C_1}	...	e_{O_1,C_k}	...	e_{O_1,C_n}
...	...	...	...	...	...
O_i	e_{O_i,C_1}	...	e_{O_i,C_k}	...	e_{O_i,C_n}
...	...	...	...	...	...
O_m	e_{O_m,C_1}	...	e_{O_m,C_k}	...	e_{O_m,C_n}

In the next step of the ICrA, relations are formed between every two elements of the matrix; thus, the evaluations of the criteria are compared in pairs for all objects in the matrix (Figure 6).

The relation $R(e_{O_i,C_k}, e_{O_j,C_k})$ has dual relation $\overline{R}$, which is true in the cases when relation R is false and vice versa [16]. Two intuitionistic fuzzy counters, $S^{\mu}_{k,l}$ and $S^{\nu}_{k,l}$, are formed and are incremented based on the following rules:

- $S^{\mu}_{k,l}$ is the number of cases in which the relations $R(e_{O_i,C_k}, e_{O_j,C_k})$ and $R(e_{O_i,C_l}, e_{O_j,C_l})$ (or the relations $\overline{R}(e_{O_i,C_k}, e_{O_j,C_k})$ and $\overline{R}(e_{O_i,C_l}, e_{O_j,C_l})$) are simultaneously satisfied.
- $S^{\nu}_{k,l}$ is the number of cases in which the relation $R(e_{O_i,C_k}, e_{O_j,C_k})$ and $\overline{R}(e_{O_i,C_l}, e_{O_j,C_l})$ (or the relations $\overline{R}(e_{O_i,C_k}, e_{O_j,C_k})$ and $R(e_{O_i,C_l}, e_{O_j,C_l})$) are simultaneously satisfied.

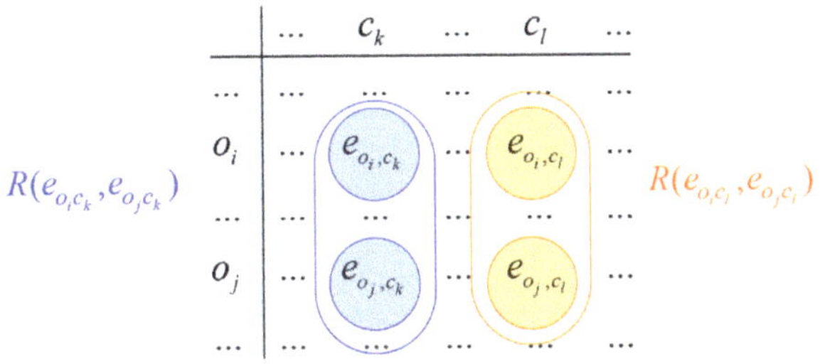

Figure 6. Formation of relations in the ICrA.

The total number of pairwise comparisons between the evaluations of m objects is $m(m-1)/2$; therefore, $0 \leq S^{\mu}_{k,l} + S^{\nu}_{k,l} \leq \frac{m(m-1)}{2}$.

For every k, l ($1 \leq k \leq l \leq m$ and $m \geq 2$), two normalized values are obtained from the above counters:

- $\mu_{C_k,C_l} = 2\frac{S^{\mu}_{k,l}}{m(m-1)}$, called the *degree of agreement* in terms of ICrA, and
- $\nu_{C_k,C_l} = 2\frac{S^{\nu}_{k,l}}{m(m-1)}$, called the *degree of disagreement* in terms of ICrA.

The pair $\langle \mu_{C_k,C_l}, \nu_{C_k,C_l} \rangle$ constructed from these two numbers plays the role of intuitionistic fuzzy evaluation of the relation between any two criteria, C_k and C_l. The output IM puts together all these intuitionistic fuzzy pairs as follows:

	C_1	...	C_k	...	C_n
C_1	$\langle 1, 0 \rangle$	...	$\langle \mu_{C_1,C_k}, \nu_{C_1,C_k} \rangle$	...	$\langle \mu_{C_1,C_n}, \nu_{C_1,C_n} \rangle$
...	...	...	...	...	...
C_k	$\langle \mu_{C_k,C_1}, \nu_{C_k,C_1} \rangle$	...	$\langle 1, 0 \rangle$	...	$\langle \mu_{C_k,C_n}, \nu_{C_k,C_n} \rangle$
...	...	...	...	...	...
C_n	$\langle \mu_{C_n,C_1}, \nu_{C_n,C_1} \rangle$	...	$\langle \mu_{C_n,C_k}, \nu_{C_n,C_k} \rangle$	...	$\langle 1, 0 \rangle$

The thresholds for μ_{C_k,C_l} and ν_{C_k,C_l} are determined algorithmically or are indicated by the user. Let $\alpha, \beta \in [0, 1]$ (with $\alpha > \beta$) be the threshold values; then the pair of criteria C_k and C_l are said to be in one of the following:

- *positive consonance*, whenever $\mu_{C_k,C_l} > \alpha$ and $\nu_{C_k,C_l} < \beta$;
- *negative consonance*, whenever $\mu_{C_k,C_l} < \beta$ and $\nu_{C_k,C_l} > \alpha$;
- *dissonance*, otherwise.

Further clarifications on determining *consonance* or *dissonance* can be found in [21].

3.4. Software Implementation of ICrA

A software application that implements the ICrA algorithm called ICrAData, v.2.5, was used during the calculations. It is freely available at https://intercriteria.net/software/

(accessed on 22 May 2025), along with the source code, a README file, and proper examples. The ICrAData interface allows for different variants of the ICrA and different algorithms for intercriteria relations calculations to be chosen, the input matrix to be read from a file or pasted from the clipboard, the input matrix to be transposed according to the investigation aims, the index matrices for the degrees of the agreements and disagreements to be visualized according to the user's needs, the threshold values for α and β to be defined by the user (conditionally defined in ICrAData as $\alpha = 0.75$ and $\beta = 0.25$, as introduced in [21]), the graphical interpretation of the results to be visualized by the intuitionistic fuzzy triangle, etc. For better visualization, the tables presenting index matrices for the degrees of the agreements and disagreements use colors for the cells' values: green to indicate *positive consonance*, red for *negative consonance*, and magenta for the cases of *dissonance*.

3.5. Correlation Analysis

The correlations between every two variables investigated in this study were calculated in MS Excel using Pearson's correlation coefficient (Pearson product moment correlation coefficient) [26].

4. Conclusions

Altogether, five scoring functions available in the MOE software package (London dG, Affinity dG, Alpha HB, ASE, and GBVI/WSA dG) were compared for their performance by using four docking outputs (BestDS, BestRMSD, RMSD_BestDS, and DS_BestRMSD) on the benchmark subset CASF-2013 consisting of 195 protein–ligand complexes from the PDBbind database.

The collected information was subjected to the ICrA using the ICrAData software and, additionally, to the CA. The most important outcomes from this investigation might be summarized as follows: (i) to the best of our knowledge, this is the first systematic investigation of the ICrA to assess the pairwise performance comparison of scoring functions available in a molecular software package, considering in parallel a variety of data outputs from molecular docking simulations; (ii) the analysis is also the first systematic investigation of the impact of different threshold values for the intervals of *consonance* and *dissonance* in terms of the ICrA; (iii) the scoring functions Alpha HB and London dG were outlined as a pair appearing in *positive consonance* in all investigated docking outputs, suggesting that these scoring functions might be used interchangeably; (iv) two pairs of scoring functions (Affinity dG–London dG and GBVI/WSA dG–London dG) were mostly in *dissonance* (except for BestRMSD), suggesting that they can complement each other in consensus docking studies; (v) for the docking output BestRMSD, all pairs of scoring functions were in *positive consonance*, thus confirming that the scoring functions studied perform best in reproducing the binding poses of the co-crystallized ligands; (vi) the comparison between the ICrA and CA results illustrates the ability of the ICrA compared to the CA to identify new relations between the investigated criteria.

The proposed approach can be applied to any other scoring functions and software packages as well as any other datasets of protein–ligand complexes of research interest.

Supplementary Materials: The following supporting information can be downloaded at: https://www.mdpi.com/article/10.3390/molecules30132777/s1: Table S1: Results obtained after ICrA application at $\alpha = 0.75$ and $\beta = 0.25$; Table S2: Results obtained after ICrA application at $\alpha = 0.70$ and $\beta = 0.30$; Table S3: Results obtained after ICrA application at $\alpha = 0.67$ and $\beta = 0.33$; Table S4: Results obtained after ICrA application at $\alpha = 0.65$ and $\beta = 0.35$; Table S5: Results obtained after ICrA application at $\alpha = 0.60$ and $\beta = 0.40$.

Author Contributions: Conceptualization, T.P., I.P. and K.A.; methodology, T.P., I.P. and K.A.; software, M.A., P.A., I.T., D.J., I.L. and T.P.; validation, M.A., P.A., I.T., I.L. and T.P.; formal analysis, M.A., P.A., I.T., D.J., I.L., K.A., I.P. and T.P.; investigation, M.A., P.A., I.T., D.J., I.L., K.A., I.P. and T.P.; resources, M.A., P.A., I.T., D.J. and I.L.; data curation, M.A., P.A., I.T., D.J. and I.L.; writing—original draft preparation, M.A., I.P. and T.P.; writing—review and editing, M.A., P.A., I.T., D.J., I.L., K.A., I.P. and T.P.; visualization, M.A., D.J., I.P. and T.P.; supervision, K.A., I.P. and T.P.; project administration, K.A., I.P. and T.P.; funding acquisition, I.T., K.A. and T.P. All authors have read and agreed to the published version of the manuscript.

Funding: K.A. and T.P. gratefully acknowledge the funding by the Bulgarian National Science Fund under the grant KP-06-N72/8 from 14 December 2023, "Intuitionistic Fuzzy Methods for Data Analysis with an Emphasis on the Blood Donation System in Bulgaria", while M.A., P.A., I.T., D.J., I.L. and T.P. gratefully acknowledge the grant KP-06-COST/3 from 23 May 2023 with regard to COST Action CA21145 "European Network for Diagnosis and Treatment of Antibiotic-resistant Bacterial Infections (EURESTOP)".

Institutional Review Board Statement: Not applicable.

Informed Consent Statement: Not applicable.

Data Availability Statement: The original contributions presented in the study are included in the article and Supplementary Material. Further inquiries can be directed to the corresponding author.

Conflicts of Interest: The authors declare no conflicts of interest.

References

1. Höltje, H.-D.; Sippl, W.; Rognan, D.; Folkers, G. *Molecular Modeling: Basic Principles and Applications*, 3rd rev. expanded ed.; Wiley-VCH: Weinheim, Germany, 2008.
2. Cheng, T.; Li, X.; Li, Y.; Liu, Z.; Wang, R. Comparative Assessment of Scoring Functions on a Diverse Test Set. *J. Chem. Inf. Model.* **2009**, *49*, 1079–1093. [CrossRef] [PubMed]
3. Li, Y.; Liu, Z.H.; Han, L.; Li, J.; Liu, J.; Zhao, Z.X.; Li, C.K.; Wang, R.X. Comparative Assessment of Scoring Functions on an Updated Benchmark: I. Compilation of the Test Set. *J. Chem. Inf. Model.* **2014**, *54*, 1700–1716. [CrossRef] [PubMed]
4. Xu, W.; Lucke, A.J.; Fairlie, D.P. Comparing Sixteen Scoring Functions for Predicting Biological Activities of Ligands for Protein Targets. *J. Mol. Graph. Model.* **2015**, *57*, 76–88. [CrossRef] [PubMed]
5. Su, M.; Yang, Q.; Du, Y.; Feng, G.; Liu, Z.; Li, Y.; Wang, R. Comparative Assessment of Scoring Functions: The CASF-2016 Update. *J. Chem. Inf. Model.* **2019**, *59*, 895–913. [CrossRef] [PubMed]
6. Wang, R.; Fang, X.; Lu, Y.; Wang, S. The PDBbind Database: Collection of Binding Affinities for Protein-Ligand Complexes with Known Three-Dimensional Structures. *J. Med. Chem.* **2004**, *47*, 2977–2980. [CrossRef] [PubMed]
7. Yan, Z.; Wang, J. Incorporating specificity into optimization: Evaluation of SPA using CSAR 2014 and CASF 2013 benchmarks. *J. Comput. Aided Mol. Des.* **2016**, *30*, 219–227. [CrossRef] [PubMed]
8. Kalinowsky, L.; Weber, J.; Balasupramaniam, S.; Baumann, K.; Proschak, E. A Diverse Benchmark Based on 3D Matched Molecular Pairs for Validating Scoring Functions. *ACS Omega* **2018**, *3*, 5704–5714. [CrossRef] [PubMed]
9. Molecular Operating Environment. The Chemical Computing Group, v. 2016.08. Available online: http://www.chemcomp.com (accessed on 22 May 2025).
10. Gaillard, T. Evaluation of AutoDock and AutoDock Vina on the CASF-2013 Benchmark. *J. Chem. Inf. Model.* **2018**, *58*, 1697–1706. [CrossRef] [PubMed]
11. Khamis, M.A.; Gomaa, W. Comparative assessment of machine-learning scoring functions on PDBbind 2013. *Eng. Appl. Artif. Intell.* **2015**, *45*, 136–151. [CrossRef]
12. Li, Y.; Su, M.; Han, L.; Liu, Z.; Wang, R. Comparative Assessment of Scoring Functions on an Updated Benchmark: 2. Evaluation Methods and General Results. *J. Chem. Inf. Model.* **2014**, *54*, 1717–1736. [CrossRef] [PubMed]
13. Wang, Y.; Qiu, Z.; Jiao, Q.; Chen, C.; Meng, Z.; Cui, X. Structure-based Protein-drug Affinity Prediction with Spatial Attention Mechanisms. In Proceedings of the 2021 IEEE International Conference on Bioinformatics and Biomedicine (BIBM), Houston, TX, USA, 9–12 December 2021; pp. 92–97.
14. Wang, Y.; Wei, Z.; Xi, L. Sfcnn: A novel scoring function based on 3D convolutional neural network for accurate and stable protein-ligand affinity prediction. *BMC Bioinform.* **2022**, *23*, 222. [CrossRef] [PubMed]

15. Rana, M.M.; Nguyen, D.D. Geometric graph learning with extended atom-types features for protein-ligand binding affinity prediction. *Comput. Biol. Med.* **2023**, *164*, 107250. [CrossRef] [PubMed]
16. Atanassov, K.; Mavrov, D.; Atanassova, V. Intercriteria Decision making: A new approach for multicriteria decision making, based on index matrices and intuitionistic fuzzy sets. *Issues IFSs GNs* **2014**, *11*, 1–8.
17. Chorukova, E.; Marinov, P.; Umlenski, I. Survey on Theory and Applications of InterCriteria Analysis Approach. In *Research in Computer Science in the Bulgarian Academy of Sciences. Studies in Computational Intelligence*, 1st ed.; Atanassov, K., Ed.; Springer: Cham, Switzerland, 2021; Volume 934, pp. 453–469.
18. Jereva, D.; Angelova, M.; Tsakovska, I.; Alov, P.; Pajeva, I.; Miteva, M.; Pencheva, T. InterCriteria Analysis Approach for Decision-making in Virtual Screening: Comparative Study of Various Scoring Functions. *Lect. Notes Netw. Syst.* **2022**, *374*, 67–78.
19. Jereva, D.; Pencheva, T.; Tsakovska, I.; Alov, P.; Pajeva, I. Exploring Applicability of InterCriteria Analysis to Evaluate the Performance of MOE and GOLD Scoring Functions. *Stud. Comput. Intell.* **2021**, *961*, 198–208.
20. Tsakovska, I.; Alov, P.; Ikonomov, N.; Atanassova, V.; Vassilev, P.; Roeva, O.; Jereva, D.; Atanassov, K.; Pajeva, I.; Pencheva, T. Intercriteria analysis implementation for exploration of the performance of various docking scoring functions. *Stud. Comput. Intell.* **2021**, *902*, 88–98.
21. Atanassov, K.; Atanassova, V.; Gluhchev, G. Intercriteria analysis: Ideas and problems. *Notes Intuit. Fuzzy Sets* **2015**, *21*, 81–88.
22. Medina-Franco, J.L. Grand Challenges of Computer-Aided Drug Design: The Road Ahead. *Front. Drug Discov.* **2021**, *1*, 728551. [CrossRef]
23. Atanassov, K. *Index Matrices: Towards an Augmented Matrix Calculus, Studies in Computational Intelligence*, 1st ed.; Springer International Publishing: Cham, Switzerland, 2014; Volume 573.
24. Atanassov, K. *Intuitionistic Fuzzy Logics*; Springer: Cham, Switzerland, 2017.
25. Zadeh, L.A. Fuzzy Sets. *Inf. Control.* **1965**, *8*, 338–353. [CrossRef]
26. Correlation Analysis. Available online: https://www.questionpro.com/features/correlation-analysis.html (accessed on 9 January 2025).

molecules

MDPI

Article

Generation of Rational Drug-like Molecular Structures Through a Multiple-Objective Reinforcement Learning Framework

Xiangying Zhang, Haotian Gao, Yifei Qi, Yan Li * and Renxiao Wang *

Department of Medicinal Chemistry, School of Pharmacy, Fudan University, 826 Zhangheng Road, Shanghai 201203, China
* Correspondence: li_yan@fudan.edu.cn (Y.L.); wangrx@fudan.edu.cn (R.W.)

Abstract: As an appealing approach for discovering novel leads, the key advantage of de novo drug design lies in its ability to explore a much broader dimension of chemical space, without being confined to the knowledge of existing compounds. So far, many generative models have been described in the literature, which have completely redefined the concept of de novo drug design. However, many of them lack practical value for real-world drug discovery. In this work, we have developed a graph-based generative model within a reinforcement learning framework, namely, METEOR (Molecular Exploration Through multiplE-Objective Reinforcement). The backend agent of METEOR is based on the well-established GCPN model. To ensure the overall quality of the generated molecular graphs, we implemented a set of rules to identify and exclude undesired substructures. Importantly, METEOR is designed to conduct multi-objective optimization, i.e., simultaneously optimizing binding affinity, drug-likeness, and synthetic accessibility of the generated molecules under the guidance of a special reward function. We demonstrate in a specific test case that without prior knowledge of true binders to the chosen target protein, METEOR generated molecules with superior properties compared to those in the ZINC 250k data set. In conclusion, we have demonstrated the potential of METEOR as a practical tool for generating rational drug-like molecules in the early phase of drug discovery.

Keywords: molecular generative model; de novo drug design; multi-objective optimization; GCPN

1. Introduction

Virtual screening of compound libraries has been a widely adopted approach in structure-based drug discovery for finding novel lead compounds. However, the potential exploration of molecules with desired properties is severely curtailed by the limited size of available compound libraries (~10^9) [1]. This constraint pales in comparison with the vast chemical space of "drug-like" compounds, which is estimated to range from 10^{23} to 10^{60} [2]. To bridge this gap, de novo drug design offers another approach to delving into the chemical space beyond existing compounds. Conventional de novo design methods typically rely on a pre-defined fragment library to construct molecular structures in a stepwise manner. Such a building-up process is relatively time-consuming, and yet the structural diversity among the generated molecular structures is in principle limited by the fragment library employed therein. Moreover, conventional de novo design methods often produce molecular structures that are challenging to synthesize due to extensive enumeration [3,4]. All these obstacles have hindered the wide application of de novo design to practical drug discovery efforts.

 https://doi.org/10.3390/molecules30010018

In recent years, generative models, a type of unsupervised training model, have emerged as invaluable tools in various scientific domains [5]. Such models have been able to generate new samples by comprehending the essential probability distribution underlying the given training samples. Generative models quickly found their applications in the realm of chemistry, where they were typically trained on large compound libraries to capture the intrinsic probability distribution embedded in the molecular structures. By drawing samples from the learned distribution, novel molecular structures were generated, which effectively expanded the accessible chemical space. It has been demonstrated that even a tiny fraction, for example, 0.1%, of a compound library, when used to train a generative model, could cover a significant portion of the chemical space spanned by the entire library [6]. Thus, generative models hold great promise in expanding the arsenal for drug discovery. Particularly for de novo drug design, generative models can not only create new molecules but also to craft molecules of specific interest.

Reinforcement learning presents an approach for achieving targeted molecule generation [7]. Within the framework of reinforcement learning, an agent engages with an environment through a sequence of actions. The agent iteratively refines its policy to maximize cumulative rewards across the action sequence, guided by the environment's feedback. In the context of de novo drug design employing reinforcement learning, an environment is tailored to provide rewards to the agent based on the properties of the generated molecules. Previous studies have demonstrated the utility of reinforcement learning in biasing generative models toward the creation of molecules with desired optimized properties [8–18].

However, many of the current generative models exhibit certain limitations when being evaluated in real-world drug discovery scenarios. For example, some models aim at overly contrived objectives, such as maximization of log P without any limit [8,9,12,17,19]. Some other models focus exclusively on the binding affinity against a specific target [10,11,14–16]. However, a successful drug discovery process is multi-objective in nature, where one has to consider and evaluate multiple properties of the candidates simultaneously [20–23]. Therefore, we believe that a generative model with practical value for de novo drug design has to be trained in a multi-objective manner.

Accordingly, we have developed such a molecular generative model, namely, METEOR (Molecular Exploration Through multiplE-Objective Reinforcement). METEOR is integrated with a reinforcement learning framework, which allows the rapid design of molecules with desirable drug-likeness and synthetic accessibility, as well as binding affinity to a user-defined target protein. In METEOR, we employ the Graph Convolutional Policy Network (GCPN) originally proposed by You et al. [8] as the fundamental architecture to construct the backend generative model. We evaluated several graph traversal algorithms [24] in terms of their efficiency in molecular structure generation. We also introduced chemical rules to detect improper substructures, thereby substantially elevating the quality of the molecular structures generated. Importantly, we introduced a special reward function to promote multi-objective optimization, combining considerations of binding affinity to the target protein, drug-likeness, and synthetic accessibility. Here, binding affinity to the target protein was evaluated by PLANET, a GNN-based deep learning model developed by our group [25]. Finally, we showcased the potential application of METEOR to real-world drug discovery with a retrospective example.

2. Results and Discussion

2.1. Comparison of Generative Models Based on Different Algorithms

The several generative models developed in our study were trained on the ZINC 250k data set in order to generate valid molecular graphs. To evaluate the performance of these models in this aspect, metrics encompassing validity, uniqueness, and novelty were considered. These metrics were assessed based on a sample of 50,000 molecules generated by each model.

The validity of the generated molecules remained at 100% across all models (Table 1), which should be attributed to the step-by-step valency check enabled during graph generation. In contrast to SMILES-based models, which might encounter validity issues due to syntax problems, graph-based generative models benefit from a more natural representation of molecular structures, ensuring high validity. However, a validity check by RDKit does not guarantee drug-like molecular structures (Figure S1 in the Supporting Information). Thus, we have implemented additional chemical rules in the molecular generation environment in our model to filter out undesired substructures, including cumulative alkenes and peroxyl bonds, double or triple bonds in three or four-membered rings, bridged rings formed with aromatic rings, and large rings. Our analysis indicated that these additional substructure detections led to the elimination of approximately 40% of the impractical structures generated by $GCPN_{origin}$. Moreover, the breadth-first model (BFM) was observed to have a problem with ring closure (Figure S2 in the Supporting Information). This problem arose due to the divergent nature of breadth-first graph generation, where the generative model tended to generate molecular graphs with incomplete rings, which may be closed later after several inconsecutive actions. In contrast, GCPN and the depth-first model (DFM) in principle can generate molecular graphs with rings in a more practical and complete manner.

Table 1. Metrics of 50,000 molecules generated by several generative models.

Models	Validity			Uniqueness		Novelty	
	RDKit	Pattern	Completeness	Molecule	Scaffold	Molecule	Scaffold
GCPN (origin)	1.000	0.592	0.993	1.000 [a]	0.666 [a]	1.000 [a]	0.928 [a]
GCPN (ours)	1.000	1.000	0.960	1.000	0.737	1.000	0.953
DFM	1.000	1.000	0.987	0.912	0.626	0.999	0.914
BFM	1.000	1.000	0.677	0.776	0.454	1.000	0.925

[a]: Molecules containing improper substructures are viewed as invalid.

Among our evaluation metrics, uniqueness reflects the fraction of non-duplicate molecules, while novelty reflects the fraction of generated molecules not presented in the training set. Our results show that BFM exhibits the lowest performance in terms of uniqueness and novelty (Table 1). By analyzing BFM-generated molecules, we have observed that the graph generation process is prone to terminate prematurely and produce simple and duplicate structures. In order to evaluate the scaffold uniqueness and novelty presented by the molecules in the ZINC 250k data set, we extracted the Bemis–Murcko scaffolds for all of them. Our results revealed that $GCPN_{ours}$ and DFM achieve similar metrics, while the performance of BFM is limited by its preference for molecules with simple structures.

To gain a deeper understanding of the chemical space covered by the molecules generated by these several generative models, we performed UMAP projection on the molecules generated by $GCPN_{ours}$ and DFM, as well as 50,000 molecules randomly selected from the ZINC 250k data set. Here, UMAP analysis was performed with the umap-learn Python package [26]. Molecules were represented by their the extended-connectivity fingerprints (ECPF4) fingerprints hashed to 1024 bits. The resulting binary vectors were then reduced to 250 dimensions using principal component analysis before being projected onto two dimensions. The results are illustrated in Figure 1. One can see that both the outcomes given by $GCPN_{ours}$ and DFM effectively span the chemical space represented by the training set (i.e., ZINC 250k), indicating their comparable ability to generate diverse molecule structures.

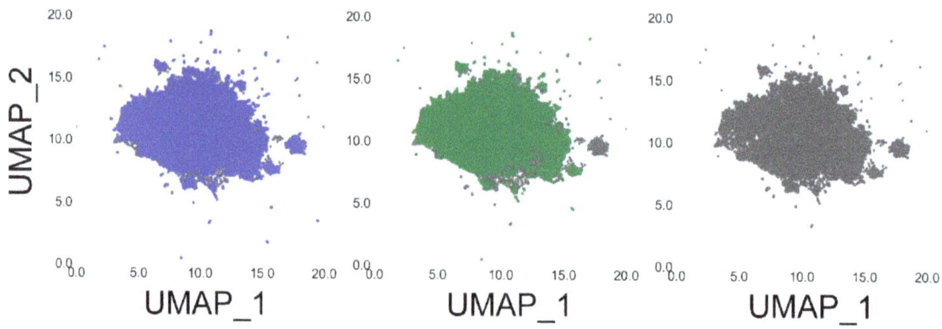

Figure 1. UMAP projection of the ZINC 250k molecules (grey) and those generated by DFM (blue) and $GCPN_{ours}$ (green), respectively. This plot illustrates the similarity between the chemical spaces covered by different generative models.

2.2. *Test Case: De Novo Design with METEOR*

As discussed above, $GCPN_{ours}$ and DFM demonstrated a remarkable advantage over BFM, this test attempted to evaluate the performance of $METEOR_{GCPN}$ and $METEOR_{DFM}$ within the realm of reinforcement learning. We then wanted to examine their performance in a real de novo drug design scenario. The objective here was to design ligand molecules targeting glucocerebrosidase (GBA), simultaneously optimizing essential properties including drug-likeness, synthetic accessibility, and binding affinity to the target.

To investigate the effect of multi-objective optimization, we examined the three desired features (i.e., drug-likeness, synthetic accessibility, and binding affinity) of the molecules generated at the initial round and the final round of reinforcement learning (Figure 2). Firstly, a notable improvement in the predicted binding affinity can be observed if comparing the molecules generated by $METEOR_{DFM}$, $METEOR_{GCPN}$, and those from ZINC 250k. Regarding the QED value, a significant fraction of the molecules generated by $METEOR_{DFM}$ and $METEOR_{GCPN}$ (77.6% and 83.8%, respectively) exceeded the QED threshold of 0.6. Nevertheless, no notable improvement in the QED value was observed after reinforcement learning. This is because the ZINC 250k data set as a whole already exhibits a high level of QED value, leaving very limited room for further improvement. Regarding the SAScore value, the majority of ZINC 250k molecules fall within the range of (1.5, 5.0). After reinforcement learning, SAScore of the generated molecules concentrated at the range of (2.5, 3.5) with a more focused distribution. Furthermore, both models generated fewer molecules that were predicted to be challenging for synthesis as compared to the ZINC 250k molecules. Considering all three features together, the distribution of the unweighted sum of three feature rewards shifted to the right as compared to the distribution of the ZINC 250k molecules. To conclude, both $METEOR_{DFM}$ and $METEOR_{GCPN}$ were able to generate novel molecules with improved predicted binding affinity under the constraints of drug-likeness and synthetic accessibility.

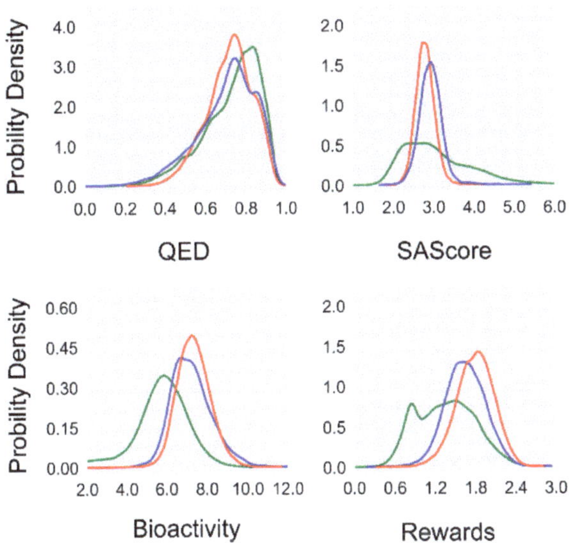

Figure 2. Distribution of three desired features and the unweighted sum of rewards of the molecules generated at the last round of reinforcement learning (red lines: $\text{METEOR}_{\text{GCPN}}$; blue lines: $\text{METEOR}_{\text{DFM}}$; green lines: ZINC 250k).

2.3. *$\text{METEOR}_{\text{GCPN}}$: Has a Larger Action Space as Well as a Higher Learning Efficiency*

Incorporating the depth-first graph traversal algorithms in DFM eliminates the need to decide the starting atom for adding a new bond. This modification reduces the action space of $\text{METEOR}_{\text{DFM}}$ and theoretically streamlines reinforcement learning. However, the learning curve demonstrated that $\text{METEOR}_{\text{GCPN}}$ can be trained at a higher level of stability and efficiency than $\text{METEOR}_{\text{DFM}}$ in reinforcement learning (Figure 3a,b). After 50 rounds of reinforcement learning, $\text{METEOR}_{\text{GCPN}}$ received a mean total reward of around 1.25, whereas the mean total reward of $\text{METEOR}_{\text{DFM}}$ at the same point was approximately 0.85. Despite the smaller action space of $\text{METEOR}_{\text{DFM}}$, the full trajectory for $\text{METEOR}_{\text{DFM}}$ for generating a molecular graph is roughly twice as long as that of $\text{METEOR}_{\text{GCPN}}$. This inequality accounts for the different efficiency of $\text{METEOR}_{\text{DFM}}$ and $\text{METEOR}_{\text{GCPN}}$. For example, over a three-day period of reinforcement learning, $\text{METEOR}_{\text{GCPN}}$ generated around 2.7 million molecules across 212 rounds, whereas $\text{METEOR}_{\text{DFM}}$ generated around 1.8 million molecules over 138 rounds. Given the same amount of training time, the additional training iterations achieved by $\text{METEOR}_{\text{GCPN}}$ make it possible to uncover molecules with improved properties. Moreover, GCPN's inherent capability of determining when to terminate the graph expansion allows $\text{METEOR}_{\text{GCPN}}$ to assess the attributes of the existing molecular structure. This capability empowers $\text{METEOR}_{\text{GCPN}}$ to judiciously halt the expansion of a graph when the current structure exhibits particularly favorable attributes. This explains why $\text{METEOR}_{\text{GCPN}}$ generated molecules with superior synthetic accessibility in comparison to $\text{METEOR}_{\text{DFM}}$.

In addition, a notable disparity was observed between the total reward and the unpenalized reward acquired by both $\text{METEOR}_{\text{GCPN}}$ and $\text{METEOR}_{\text{DFM}}$ (Figure 3c,d). This gap primarily arose from the property penalty at the early training phase (Figure 3e,f). In METEOR, property penalty (Equation (5)) was the driving force for multi-objective optimization on binding affinity to the protein, drug-likeness, and synthetic accessibility. Computing rewards by a weighted sum across three property rewards reinforced the optimization to be conducted toward all three properties.

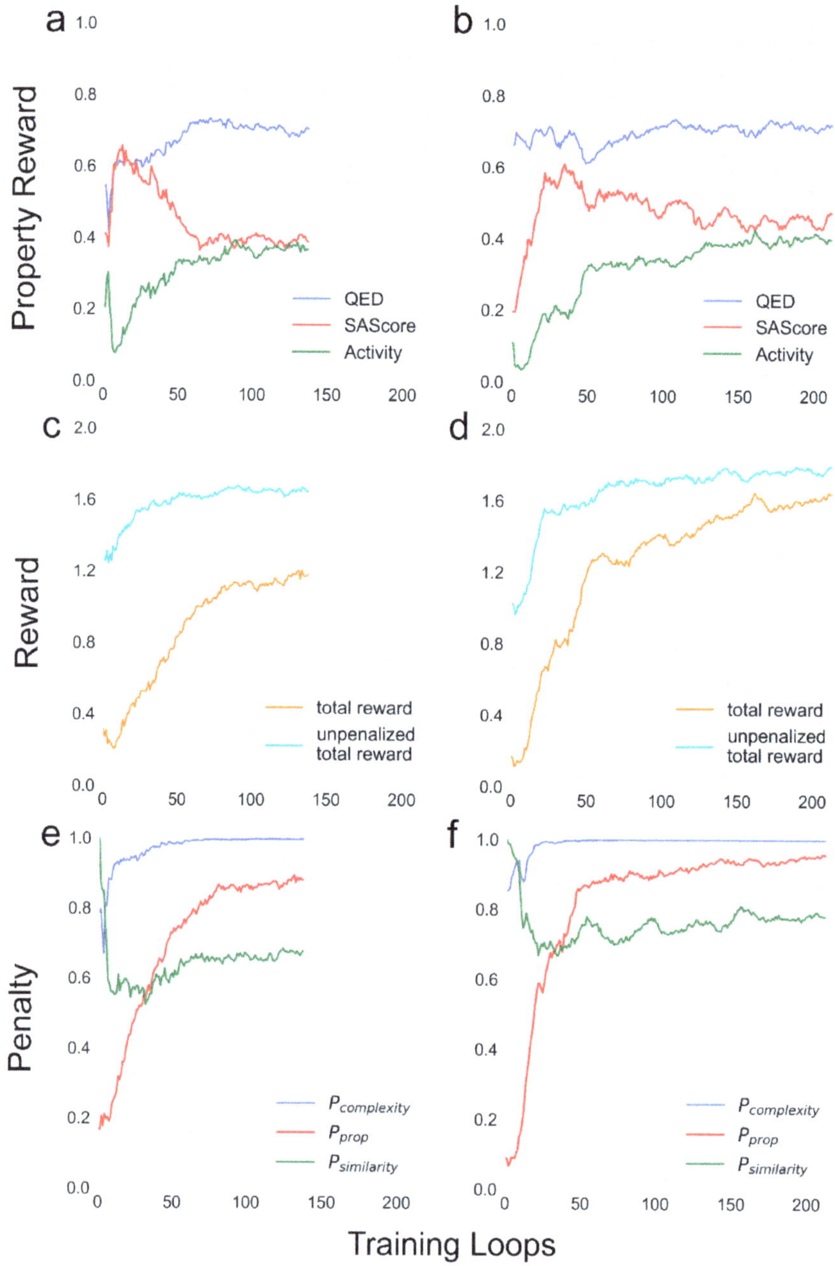

Figure 3. Several key features of METEOR$_{DFM}$ (**left**) and METEOR$_{GCPN}$ (**right**) observed during the reinforcement learning process. (**a**,**b**): Three property rewards; (**c**,**d**): Total and unpenalized rewards; (**e**,**f**): Three penalty factors. Here, each line plots the mean value of a certain feature computed over all molecules generated at each round of roll-out.

The complexity penalty was computed primarily by counting heavy atoms. This penalty was introduced to balance the bias along structure generation, where larger molecules tend to receive higher predicted binding scores by PLANET. Moreover, larger molecules often contain challenging moieties for chemical synthesis, such as chiral centers. The similarity penalty was introduced to encourage METEOR to explore the chemical space preventing it from becoming confined to local maxima. During the initial training rounds, a relatively modest similarity penalty was observed among the molecules generated by both $METEOR_{DFM}$ and $METEOR_{GCPN}$ due to the presence of limited high-scoring molecules recorded in the memory stack. At the 40th round or so, the influence of similarity penalties became more obvious (Figure 3e,f). Here, both $METEOR_{DFM}$ and $METEOR_{GCPN}$ were able to explore the full chemical space covered by the ZINC 250k data set throughout the training process without being trapped in certain restricted regions (Figure 4).

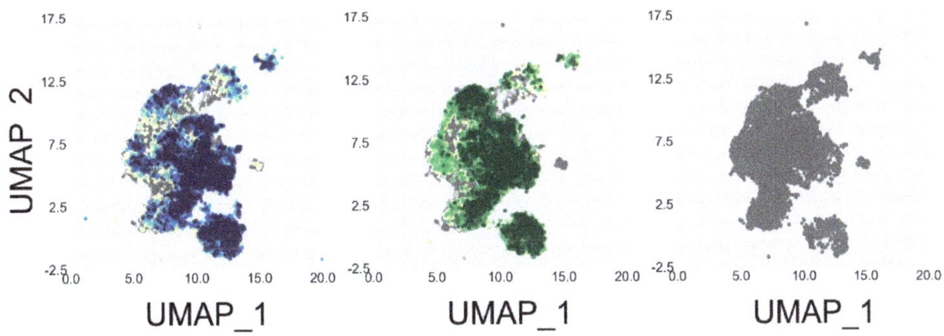

Figure 4. UMAP projections of 50,000 ZINC 250k molecules (grey) and 50,000 molecules generated at certain rounds by $METEOR_{DFM}$ (blue) and $METEOR_{GCPN}$ (green), respectively. Different rounds are indicated in colors with different shades.

In conclusion, both $METEOR_{DFM}$ and $METEOR_{GCPN}$ are able to generate molecules with optimized properties. The major distinction between $METEOR_{DFM}$ and $METEOR_{GCPN}$ lies in their efficiency.

2.4. The Practical Value of METEOR in De Novo Drug Design

In this study, we evaluated the practical value of METEOR in de novo drug design by using GBA as a test case. The quality of the molecules generated by METEOR was reflected by analyzing their similarity to true binders of GBA collected from ChEMBL. If using an ECFP4 Tanimoto coefficient of 0.6 as the threshold, 15 molecules generated by $METEOR_{DFM}$ shared similar structures to true binders to GBA. As for $METEOR_{GCPN}$, this number was 17. A few such examples are given in Figure 5. One can see that the generated molecules shared an almost identical scaffold as a certain GBA binder. This observation demonstrated that METEOR is able to generate drug-like molecules with potential value.

It should be mentioned though that as a whole, a substantial proportion of the true GBA binders considered in our study have a QED value below 0.5 or SAScores over 3.5 (Figure S3 in the Supporting Information). However, the majority of the molecules generated by METEOR had optimized QED values and SAScores that do not stay at this range (Figure 2). Thus, in this particular test cast, this gap resulted in rather limited matched pairs between the outcomes of METEOR and true GBA binders.

Moreover, we employed the GLIDE module in the Schrödinger software, a widely used conventional molecule docking method, to evaluate the binding affinity of the molecules generated by METEOR with the target protein. Prior to the molecular docking job, the molecules generated during the reinforcement learning process in METEOR were filtered based on the following criteria: (1) an QED value above 0.6, (2) an SAScore lower than 3.0, and (3) a predicted binding affinity value greater than 7.0 (in -log units). The molecules meeting all requirements were then docked into the binding pocket on GBA. To make a comparison, all ZINC 250k molecules were also docked into the binding pocket on GBA through the same protocol.

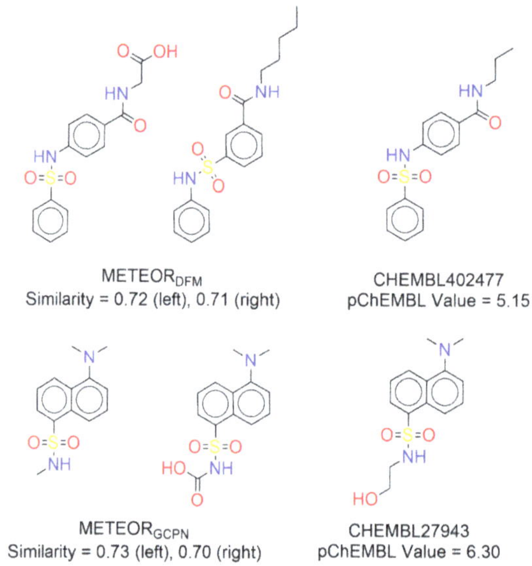

Figure 5. Examples of the molecules generated by METEOR$_{DFM}$ and METEOR$_{GCPN}$ as well as the corresponding true binders of GBA.

As shown in Figure 6, the molecules generated by either METEOR$_{GCPN}$ or METEOR$_{DFM}$ on average had better GLIDE binding scores than those ZINC 250k molecules, even though even though the optimization of binding affinity in METEOR was guided by a different scoring function PLANET. The 1% percentile of docking scores was −6.20, −6.83, and −6.57 for molecules from ZINC 250k, METEOR$_{DFM}$, and METEOR$_{GCPN}$, respectively. Note that besides binding affinity to the target protein, the molecules generated by METEOR were also optimized in terms of drug-likeness and synthetic accessibility. Therefore, it is reasonable to expect that more promising active hits can be discovered through application of METEOR rather than a conventional virtual screening of the ZINK 250k data set.

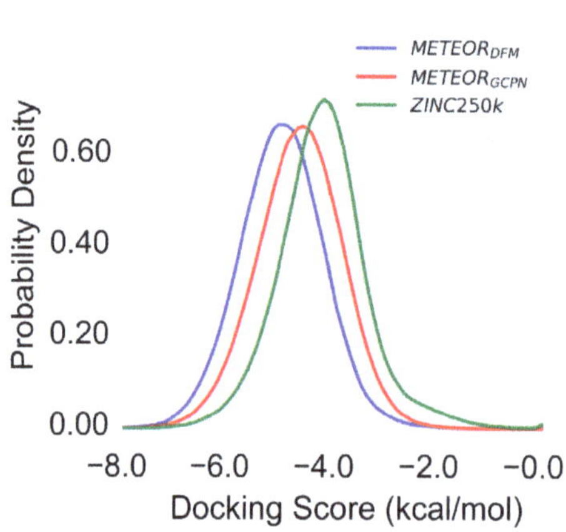

Figure 6. Distribution of the GLIDE docking scores computed for several sets of molecules: ZINC 250k molecules (n = 247168, green), METEOR$_{DFM}$ molecules (n = 279910, blue), and METEOR$_{GCPN}$ molecules (n = 739130, red).

3. Methods

3.1. The Backend Molecule Generative Models

The GCPN model extends an existing molecular graph by adding new chemical bonds one after another (Figure 7). During this process, four decisions need to be made at each step: (1) determining the starting atom ("focus atom") to which the new bond is added, (2) selecting the end atom of the new bond, (3) specifying the type of the new bond, and (4) deciding whether to terminate graph expansion [8]. The first decision significantly expands the action space of GCPN, leading to numerous possibilities within each existing subgraph. To address this complexity, we introduced two models: DFM and BFM, each employing a distinct graph traversal algorithm (Figure 7). In both models, the "focus atom", defined by the respective graph traversal algorithm, serves as the starting point for adding a new bond. In alignment with DFM and BFM, the final task in GCPN, i.e., determining whether to terminate graph generation, is replaced by marking the current focus atom as "finished". Graph generation terminates when all nodes have been marked.

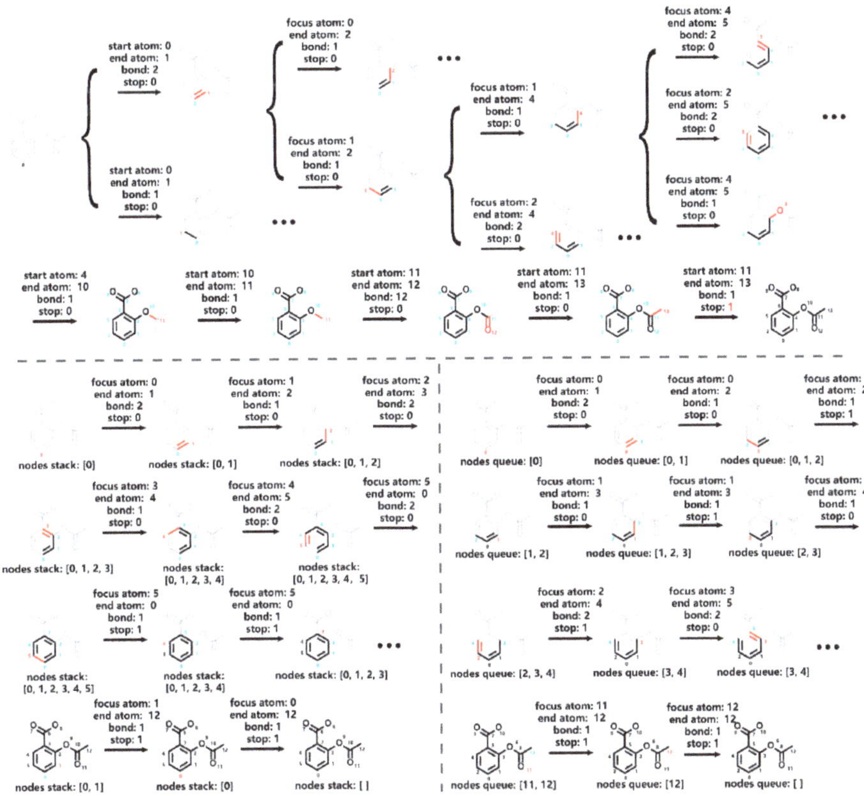

Figure 7. Illustration of graph generation process for the aspirin molecule by GCPN (**top**), DFM (**bottom left**, in a depth-first manner), and BFM (**bottom right**, in a breadth-first manner). At each step, the end atoms of new bonds are marked in red, possible focus atoms are marked in blue, and "finished atoms" are depicted in black.

Individual atom nodes were encoded using vectors with a dimension of 20, consisting of one-hot encoded element type, atom degree, and membership in rings of varying sizes from 3 to 7. These initial vectors were then embedded into a latent space (h_0) with a size of 64 dimensions. To extract features from the input graphs, we utilized the graph convolution network (GCN) architecture [27–29]. The entire molecular graph was partitioned into three distinct subgraphs based on different bond orders. One module with three separate graph convolution layers with a hidden size of 64, each with learnable parameters W_i^l, were employed on the three subgraphs, as denoted in Equation (1).

$$h^l = \sum_{i=1}^{b} \left(ReLU \left(\tilde{D}_i^{-\frac{1}{2}} \tilde{A}_i \tilde{D}_i^{-\frac{1}{2}} h^{(l-1)} W_i^l \right) \right), b \in \{1,2,3\} \quad (1)$$

A_i is the ith slice bond-conditioned adjacent matrix, $\tilde{A}_i = A_i + I$; $\tilde{D}_i$ is the ith slice bond-conditioned degree matrix with self-loop.

In our implementation, we utilized three such modules to extract the underlying features from the molecular graphs. The extracted latent features were subsequently utilized to make informed decisions within the model. The configuration of task layers in GCPN remained consistent with the original literature [8]. In the case of DFM and BFM, when selecting ending atom for the chemical bond to be added, nodes marked as "finished" were overlooked. The probability of each action $P(a_t)$ was calculated as shown in Equation (2):

$$P(a_t) = \left(1 - \mathbb{I}\left(a_t^{\text{finish}}\right)\right) \prod_j P\left(a_t^j\right) + \mathbb{I}\left(a_t^{\text{finish}}\right) P\left(a_t^{\text{finish}}\right) \quad (2)$$

Each action step a_t is composed of three sub-tasks a_t^j, i.e., selecting the end atom of the new bond, specifying type of the new bond, and whether to mark the focus atom as "finished". $\mathbb{I}(a_t^{\text{finish}})$ equals 0 if the focus atom is not decided to be marked, else $\mathbb{I}(a_t^{\text{finish}})$ equals 1.

To enhance stability and performance in reinforcement learning, a commonly employed strategy involves pre-training a generative model using an established compound database [30]. In our study, we employed the widely-used ZINC 250k data set, comprising structurally diverse "drug-like" molecules that have been synthesized in reality. Structures of these molecules were examined to eliminate those containing rings with eight or more members. The remaining molecules were considered as the ground truth and served as expert training data. For GCPN pre-training, a randomly sampled connected subgraph G′ from a molecule graph G was viewed as the state s_t. Any action a_t added an atom or a bond in G but not in G′ could be viewed as an expert action during the trajectory of generating a ground-truth molecule. The training objective was to maximize the possibility $P(a_t)$ of GCPN to take expert action a_t at state s_t. This training approach was similar to that previously reported [8]. For both DFM and BFM, the molecular structures from the filtered ZINC 250k data set were transformed into expert trajectories by randomly selecting a starting node and traversing the graphs in a depth-first or breadth-first manner, respectively. The resulting expert actions a_t consisting of the trajectories were collected for pre-training DFM and BFM. The objective in expert training for all models can be expressed as shown in Equation (3):

$$L^{\text{expert}}(\theta) = -\frac{1}{T} \sum_{t}^{T} \log P(a_t) \quad (3)$$

The Adam optimizer with a learning rate of 0.0001 was applied. After 1,000,000 training steps, GCPN, DFM, and BFM with converged loss were obtained as pre-trained generative models.

3.2. The Molecule Generation Environment

Within the context of molecular graph generation under a reinforcement learning framework, the molecule generation environment plays two essential roles, i.e., state transition dynamics and reward assignment.

3.2.1. State Transition Dynamics

The molecule generation environment plays a pivotal role in executing the actions taken by the agent, ensuring adherence to specified rules. One fundamental rule incorporated into the environment is the valency check, preventing actions that exceed an atom's maximal valency [8]. It is noteworthy that substructures adhering to the basic valency rule may not be "drug-like". Therefore, the environment in our model detects and then filters out the following "non-drug-like" substructures: (a) cumulative alkenes and peroxyl bonds; (b) double or triple bonds in a three- or four-membered ring; (c) bridged ring formed with aromatic rings; and (d) large rings with eight or more members, as detected in the smallest set of smallest rings in a molecular graph. Substructure detection is performed after each agent action through SMARTS matching (see Figure S4 in the Supporting Information). Only actions that pass both the valency check and substructure examination will be adopted by the environment to update the current molecule subgraph. Note that implementation of the above chemical rules reflects the knowledge of "drug-likeness" accumulated in the literature [31–33]. There are of course different perceptions of "drug-likeness", but the several rules listed above are relatively straightforward to be encoded in a computer program. In particular, macrocyclic structures are not allowed in our model, although some marketed drugs do consist of such structures [34]. From a practical view, macrocyclic structures are normally introduced at the stage of lead optimization to impose conformational constraints. Considering that our model will be employed primarily as an "idea generator" at the stage of lead discovery, ignoring macrocyclic structures is an acceptable trade-off for the sake of technical convenience.

3.2.2. Reward Assignment

The behavior of agents is steered by the rewards from the molecule generation environment, which can be categorized into two components: step reward and final reward. A zero-step reward is assigned to each step, except for two specific actions: (a) When a new ring is formed, a small step reward of 0.02 is assigned to encourage ring formation. (b) When an improper action is canceled by the molecular generation environment, a step reward of −0.2 is given to discourage such actions. The step reward serves to guide the agent's behavior and reduce the occurrence of improper actions. The final reward comprises several domain-specific rewards assigned based on different properties, including drug-likeness, synthetic accessibility, and predicted bio-activity. The final reward is calculated as the weighted sum of these rewards, further adjusted by a penalty factor. Reward functions related to specific properties utilize a linear scaling function that maps values between a lower bound and an upper bound, as described in Equation (4):

$$R_{prop} = \begin{cases} 1.0 & S_{prop} \geq S_{prop}^{high}\,; \\ \frac{S_{prop}-S_{prop}^{low}}{S_{prop}^{high}-S_{prop}^{low}}, & S_{prop}^{low} < S_{prop} < S_{prop}^{high}; \\ 0.0 & S_{prop} \leq S_{prop}^{low}\,. \end{cases} \quad (4)$$

This type of function is chosen based on the assumption that it is not necessary to optimize certain properties beyond desired ranges. For example, it is not necessary to further optimize the synthetic accessibility of a molecule with an SAScore lower than 2.0 because it is already good enough at this level.

Generated molecules are evaluated by the following three properties:

(a) Drug-likeness of a molecule is assessed by the QED index originally proposed by Hopkins et al. [33]. This index has a range (0.0, 1.0).

(b) Synthetic accessibility of a molecule is evaluated by SAScore, which has a range [1.0, 10.0]. SAScore is a rule-based tool for estimating synthetic accessibility, and its output is determined by the summation of fragment scores and a complexity penalty [35].

(c) Binding affinity to the target protein is predicted by PLANET, a graph neural network model developed in our group [25]. PLANET operates on two-dimensional molecular graphs as inputs and thus skips the exhaust molecular docking process. Its ultra-fast speed is suitable for processing generated molecules in a large number.

A penalty factor is also implemented to influence the agent model's behavior by scaling the sum of property rewards. This factor is determined based on three aspects:

(a) Complexity penalty ($P_{\text{complexity}}$). The complexity penalty is assigned based on the number of heavy atoms in the designed molecule, defined as a linear scaling function akin to Equation (4). The lower and upper bounds for the number of heavy atoms are set to 10 and 40, respectively. Additionally, for molecules with more than two chiral centers, a penalty factor of 0.5 will multiply $P_{\text{complexity}}$.

(b) Property penalty (P_{prop}). Since the reward is the sum of three property rewards, it is possible for an agent to receive a high reward from a molecule that possesses two excellent properties but one extremely poor property. The property penalty is applied as follows (Equation (5)):

$$P_{\text{prop}} = \prod\nolimits_{prop} \min(1.0, R_{prop}/0.2) \tag{5}$$

(c) Similarity penalty ($P_{\text{similarity}}$). Agents trained in reinforcement learning tend to generate highly-scored molecules. However, once a local maximum is reached, agents often struggle to explore other areas, leading to a phenomenon known as "policy collapse". Inspired by the work of Blaschke et al. [14], we devised a similarity penalty to encourage agents not only to focus on specific favorable regions in the chemical space yielding high scores but also to explore various areas within the space. Our algorithm for calculating the similarity penalty differs from that of Blaschke though. For example, all halogen atoms are ignored here to prevent our model from generating molecular structures with differences merely in the number and position of halogen atoms. This is important since at the stage of lead discovery, sufficient diversity in the structural scaffold is much desired, where terminal halogen atoms are not part of a structural scaffold. In fact, halogen atoms are often added to optimize bioactivity at a later stage of drug discovery. Subsequently, a mapping between the current molecule and those generated before the preceding twenty rounds of roll-out is performed. If a successful mapping is found, a zero-penalty factor is assigned. Molecules passing this mapping step proceed to subsequent similarity calculation. A stack is used to retain favorable molecules, that is, those generated over the preceding 10 rounds of roll-out with a final property reward surpassing 70% of the possible maximum. Tanimoto similarity coefficients between the ECFP4 of the transformed molecule and all stored high-quality molecules are calculated. $P_{\text{similarity}}$ is determined based on the maximal Tanimoto similarity coefficient, as outlined in Equation (6):

$$P_{\text{similarity}} \begin{cases} 0.0 & \text{Success Mapping } \textbf{OR} \text{ Tanimoto} \geq 0.7\,; \\ 1 - (\text{Tanimoto} - 0.4)/0.3, & 0.4 < \text{Tanimoto} < 0.7\,; \\ 1.0 & \text{Tanimoto} \leq 0.4\,. \end{cases} \tag{6}$$

The final reward (R_{final}) is the weighted sum of all property rewards scaled by overall penalty (Equation (7)):

$$R_{\text{final}} = \sum^{i} \alpha_i R_i \times \prod^{j} P_j \tag{7}$$

Here, i and j denote for different types of molecular properties and penalty factors, respectively.

3.2.3. Reinforcement Learning

Policy gradient-based methods are widely adopted in reinforcement learning. In our model, Proximal Policy Optimization (PPO) is adopted [36]. The learning objective can be written as Equation (8):

$$L_{PPO}^{\theta^k}(\theta) = -\sum_{(s_t,a_t)} min\left(\frac{P_\theta(a_t|s_t)}{P_\theta^k(a_t|s_t)}A^{\theta^k}(a_t,s_t), clip\left(\frac{P_\theta(a_t|s_t)}{P_\theta^k(a_t|s_t)}, 1-\varepsilon, 1+\varepsilon\right)A^{\theta^k}(a_t,s_t)\right)$$

$$A^{\theta^k}(a_t,s_t) = \sum_{t=t'}^{T_n} \gamma^{T_n - t'} R_{\text{final}} + R_{\text{step}} - b \quad (8)$$

In the objective function, γ represents the discount factor, and its value is experimentally set to 0.98. The clip value, denoted as ε, is set to 0.1. The superscript k denotes for the generative model obtained after the last round of training. The estimated advantage function $A^{\theta^k}(a_t,s_t)$ incorporates a learnable value function b. The value function takes the same molecular graph embedding obtained from the GCN layers and maps it to a scalar representing the estimated expected reward. The probability of an action taken by the generative model with parameter θ under state s_t, denoted as $P_\theta(a_t|s_t)$, is calculated using Equation (2).

3.3. Performance Evaluation

3.3.1. Evaluation in Terms of Generating Valid Molecular Structures

We assessed the performance of several pre-trained generative models, including the $\text{GCPN}_{\text{origin}}$ (only enabling valency check during graph generation), the $\text{GCPN}_{\text{ours}}$, DFM, and BFM (all three utilizing full substructure check, including valency check and improper substructure detection, see Figure S4 in the Supporting Information). Each pre-trained model was assigned the task of generating 50,000 molecules for evaluating validity, uniqueness, and novelty as follows:

$$\text{Validity} = \frac{\text{Number of valid graphs}}{\text{Number of generated graphs}}$$

$$\text{Uniqueness} = \frac{\text{Number of unique and valid graphs}}{\text{Number of valid graphs}}$$

$$\text{Novelty} = \frac{\text{Number of unique and valid graphs not in the training set}}{\text{Number of unique and valid graphs}}$$

Valid graphs were typically measured with respect to valency and bonds using RDKit's molecular structure parser.

3.3.2. Evaluation in Terms of Generating Useful Hits on a Specific Target Protein

To evaluate the effectiveness of METEOR in generating useful hits in a de novo drug design scenario, we chose GBA as the target protein, which is included in the popular LIT-PCBA benchmark for testing virtual screening methods [37]. The crystal structure of GBA used in our test is obtained from the Protein Data Bank (PDB entry 2V3D) [38]. Two pre-trained generative models, namely, $\text{GCPN}_{\text{ours}}$ and DFM, served as backends of METEOR (denoted as $\text{METEOR}_{\text{GCPN}}$ and $\text{METEOR}_{\text{DFM}}$, respectively, hereafter). Major adjustable parameters in these models are in Equation (4), where the lower bound of QED, SAScore, and binding affinity was set to 0.2, 3.5, and 5.5, respectively, and the upper bound was set to 0.8, 2.0, and 8.5, respectively.

Our test was performed on a server equipped with two NVIDIA GeForce 2080Ti GPU cards (11 GB memory), two Intel(R) Xeon(R) Silver 4210 CPUs @ 2.20 GHz, and 128 GB of

RAM. After three days of reinforcement learning with 10 parallel processes, all generated molecules were assessed in two aspects: Firstly, a total of 452 true binders of GBA were curated from ChEMBL, which were identified by a "Target ChEMBL ID" of CHEMBL2179 and a "pChEMBL value" greater than 5.0 (for example, K_d or Ki value < 10 μM). Pairs of molecules with an ECFP4 Tanimoto coefficient over 0.6 were defined as similar. The total number of molecules generated by METEOR that were similar to true binders to GBA was counted and analyzed. Secondly, the molecules generated by METEOR, filtered based on QED value, SAScore, and binding affinity, were docked into the binding pocket of GBA by using GLIDE in the standard precision (SP) mode in the Schrödinger software. To make a comparison, the molecules in the ZINC 250k data set were docked into the binding pocket of GBA following the same protocol.

4. Conclusions

In this work, we have developed a deep learning mode, called METEOR, for potential application in de novo drug design. Compared to many other generative models already described in the literature, METEOR has several distinct technical features.

Firstly, the backend agent of METEOR is based on the well-established GCPN model. We have evaluated several graph traversal algorithms within a reinforcement learning framework. Our findings indicate that depth-first graph generation (DFM) outperforms breadth-first graph generation (BFM). Its outcomes closely align with those of the original GCPN model in terms of validity, uniqueness, and novelty. This observation supports the potential value of both $METEOR_{GCPN}$ and $METEOR_{DFM}$ in de novo drug design. As demonstrated in the test case of GBA, without prior knowledge of true binders, both models are able to generate molecules with superior properties compared to those in the ZINC 250k data set.

Secondly, in order to ensure the overall validity of the generated molecular structures, we have implemented a set of chemical rules in METEOR to eliminate undesired substructures. In fact, if these rules are not enabled, a significant portion (~40%) of the generated molecule structures would be undesirable. This demonstrates the importance of integrating chemical knowledge into molecular structure generation, which has become a new trend in this field (for example, see a new generative model published recently [39]).

Last, and very importantly, unlike many other generative models that focus on a single objective, METEOR is designed to generate molecules with optimized traits regarding binding affinity, drug-likeness, and synthetic accessibility. These several properties are all indispensable for a successful candidate in the early phase of drug discovery. This makes METEOR better suited for practical applications to drug discovery.

Supplementary Materials: The following supporting information can be downloaded at: https://www.mdpi.com/article/10.3390/molecules30010018/s1, Figure S1: Examples of valid but unwanted molecular structures; Figure S2: Examples of molecules with "ring closure" issue; Figure S3: Distribution of QED, SAScore and PLANET binding scores of the known GBA binders collected from ChEMBL; Figure S4: Valency check and improper substructure detection implemented in different generative models.

Author Contributions: Conceptualization, R.W., Y.L. and X.Z.; methodology, X.Z.; software, X.Z.; validation, X.Z. and H.G.; formal analysis, X.Z.; investigation, X.Z.; resources, R.W. and Y.L.; data curation, X.Z.; writing—original draft preparation, X.Z.; writing—review and editing, R.W., Y.L., Y.Q., X.Z. and H.G.; visualization, X.Z.; supervision, R.W., Y.L. and Y.Q.; project administration, R.W. and Y.L.; funding acquisition, R.W. and Y.L. All authors have read and agreed to the published version of the manuscript.

Funding: This study was financially supported by the National Natural Science Foundation of China (Grant No. 81725022 and 82173739 to R. Wang), the Ministry of Science and Technology of China (National Key Research Program, Grant No.2023YFF1205102 to Y. Li), and the Shanghai Natural Science Foundation (Grant No. 24ZR1413800 to Y. Li).

Institutional Review Board Statement: Not applicable.

Informed Consent Statement: Not applicable.

Data Availability Statement: Source code of METEOR, as well as user manual and demo examples, are available at https://github.com/ComputArtCMCG/METEOR (accessed on 12 December 2024).

Conflicts of Interest: The authors declare no conflict of interest.

References

1. Sterling, T.; Irwin, J.J. ZINC 15—Ligand discovery for everyone. *J. Chem. Inf. Model.* **2015**, *55*, 2324–2337. [CrossRef] [PubMed]
2. Kirkpatrick, P.; Ellis, C. Chemical space. *Nature* **2004**, *432*, 823. [CrossRef]
3. Vanhaelen, Q.; Lin, Y.-C.; Zhavoronkov, A. The advent of generative chemistry. *ACS Med. Chem. Lett.* **2020**, *11*, 1496–1505. [CrossRef] [PubMed]
4. Sousa, T.; Correia, J.; Pereira, V.; Rocha, M. Generative deep learning for targeted compound design. *J. Chem. Inf. Model.* **2021**, *61*, 5343–5361. [CrossRef]
5. Atance, S.R.; Diez, J.V.; Engkvist, O.; Olsson, S.; Mercado, R. De novo drug design using reinforcement learning with graph-based deep generative models. *J. Chem. Inf. Model.* **2022**, *62*, 4863–4872. [CrossRef]
6. Arus-Pous, J.; Johansson, S.V.; Prykhodko, O.; Bjerrum, E.J.; Tyrchan, C.; Reymond, J.L.; Chen, H.; Engkvist, O. Randomized SMILES strings improve the quality of molecular generative models. *J. Cheminform.* **2019**, *11*, 71. [CrossRef]
7. Sridharan, B.; Goel, M.; Priyakumar, U.D. Modern machine learning for tackling inverse problems in chemistry: Molecular design to realization. *Chem. Commun.* **2022**, *58*, 5316–5331. [CrossRef]
8. You, J.X.; Liu, B.W.; Ying, R.; Pande, V.; Leskovec, J. Graph convolutional policy network for goal-directed molecular graph generation. *Adv. Neural Inf. Process. Syst.* **2018**, *31*, 6410–6421.
9. Mariya, P.; Mykhailo, S.; Junier, O.; Olexandr, I. MolecularRNN: Generating realistic molecular graphs with optimized properties. *arXiv* **2019**. [CrossRef]
10. Papadopoulos, K.; Giblin, K.A.; Janet, J.P.; Patronov, A.; Engkvist, O. De novo design with deep generative models based on 3D similarity scoring. *Bioorganic Med. Chem.* **2021**, *44*, 116308. [CrossRef]
11. Korshunova, M.; Huang, N.; Capuzzi, S.; Radchenko, D.S.; Savych, O.; Moroz, Y.S.; Wells, C.I.; Willson, T.M.; Tropsha, A.; Isayev, O. Generative and reinforcement learning approaches for the automated de novo design of bioactive compounds. *Commun. Chem.* **2022**, *5*, 129. [CrossRef] [PubMed]
12. Sicho, M.; Luukkonen, S.; van den Maagdenberg, H.W.; Schoenmaker, L.; Beiquignon, O.J.M.; van Westen, G.J.P. DrugEx: Deep learning models and tools for exploration of drug-like chemical space. *J. Chem. Inf. Model.* **2023**, *63*, 3629–3636. [CrossRef] [PubMed]
13. Zhang, W.; Zhang, K.; Huang, J. A simple way to incorporate target structural information in molecular generative models. *J. Chem. Inf. Model.* **2023**, *63*, 3719–3730. [CrossRef] [PubMed]
14. Blaschke, T.; Engkvist, O.; Bajorath, J.; Chen, H.M. Memory-assisted reinforcement learning for diverse molecular de novo design. *J. Cheminform.* **2020**, *12*, 68. [CrossRef]
15. Olivecrona, M.; Blaschke, T.; Engkvist, O.; Chen, H.M. Molecular de-novo design through deep reinforcement learning. *J. Cheminform.* **2017**, *9*, 48. [CrossRef]
16. Zhavoronkov, A.; Ivanenkov, Y.A.; Aliper, A.; Veselov, M.S.; Aladinskiy, V.A.; Aladinskaya, A.V.; Terentiev, V.A.; Polykovskiy, D.A.; Kuznetsov, M.D.; Asadulaev, A.; et al. Deep learning enables rapid identification of potent DDR1 kinase inhibitors. *Nat. Biotechnol.* **2019**, *37*, 1038–1040. [CrossRef]
17. Mokaya, M.; Imrie, F.; van Hoorn, W.P.; Kalisz, A.; Bradley, A.R.; Deane, C.M. Testing the limits of SMILES-based de novo molecular generation with curriculum and deep reinforcement learning. *Nat. Mach. Intell.* **2023**, *5*, 386–394. [CrossRef]
18. Mazuz, E.; Shtar, G.; Shapira, B.; Rokach, L. Molecule generation using transformers and policy gradient reinforcement learning. *Sci. Rep.* **2023**, *13*, 8799. [CrossRef]
19. Bilodeau, C.; Jin, W.; Jaakkola, T.; Barzilay, R.; Jensen, K.F. Generative models for molecular discovery: Recent advances and challenges. *WIREs Comput. Mol. Sci.* **2022**, *12*, e1608. [CrossRef]
20. Tan, Y.; Dai, L.; Huang, W.; Guo, Y.; Zheng, S.; Lei, J.; Chen, H.; Yang, Y. DRlinker: Deep Reinforcement Learning for Optimization in Fragment Linking Design. *J. Chem. Inf. Model.* **2022**, *62*, 5907–5917. [CrossRef]
21. Wengong, J.; Regina, B.; Tommi, J. Multi-Objective Molecule Generation using Interpretable Substructures. *arXiv* **2020**. [CrossRef]

22. Janet, J.P.; Ramesh, S.; Duan, C.; Kulik, H.J. Accurate Multiobjective Design in a Space of Millions of Transition Metal Complexes with Neural-Network-Driven Efficient Global Optimization. *ACS Cent. Sci.* **2020**, *6*, 513–524. [CrossRef] [PubMed]
23. Kneiding, H.; Nova, A.; Balcells, D. Directional multiobjective optimization of metal complexes at the billion-system scale. *Nat. Comput. Sci.* **2024**, *4*, 263–273. [CrossRef] [PubMed]
24. Mercado, R.; Bjerrum, E.J.; Engkvist, O. Exploring graph traversal algorithms in graph-based molecular generation. *J. Chem. Inf. Model.* **2022**, *62*, 2093–2100. [CrossRef] [PubMed]
25. Zhang, X.; Gao, H.; Wang, H.; Chen, Z.; Zhang, Z.; Chen, X.; Li, Y.; Qi, Y.; Wang, R. PLANET: A multi-objective graph neural network model for protein–ligand binding affinity prediction. *J. Chem. Inf. Model.* **2023**, *64*, 2205–2220. [CrossRef]
26. McInnes, L.; Healy, J.; Melville, J. UMAP: Uniform manifold approximation and projection for dimension reduction. *arXiv* **2018**, arXiv:1802.03426.
27. Duvenaud, D.; Maclaurin, D.; Aguilera-Iparraguirre, J.; Gómez-Bombarelli, R.; Hirzel, T.; Aspuru-Guzik, A.; Adams, R.P. Convolutional networks on graphs for learning molecular fingerprints. *Adv. Neural Inf. Process. Syst.* **2015**, *2*, 2224–2232.
28. Kearnes, S.; McCloskey, K.; Berndl, M.; Pande, V.; Riley, P. Molecular graph convolutions: Moving beyond fingerprints. *J. Comput. Aided Mol. Des.* **2016**, *30*, 595–608. [CrossRef]
29. Kipf, T.N.; Welling, M. Semi-supervised classification with graph convolutional networks. *arXiv* **2016**. [CrossRef]
30. Levine, S.; Koltun, V. Guided policy search. *Proc. Mach. Learn.* **2013**, *28*, 1–9.
31. Leeson, P.; Springthorpe, B. The influence of drug-like concepts on decision-making in medicinal chemistry. *Nat. Rev. Drug Discov.* **2007**, *6*, 881–890. [CrossRef] [PubMed]
32. Ursu, O.; Rayan, A.; Goldblum, A.; Oprea, T.I. Understanding drug-likeness. *WIREs Comput. Mol. Sci.* **2011**, *1*, 760–781. [CrossRef]
33. Bickerton, G.R.; Paolini, G.V.; Besnard, J.; Muresan, S.; Hopkins, A.L. Quantifying the chemical beauty of drugs. *Nat. Chem.* **2012**, *4*, 90–98. [CrossRef]
34. Jimenez, D.G.; Poongavanam, V.; Kihlberg, J. Macrocycles in Drug Discovery: Learning from the Past for the Future. *J. Med. Chem.* **2023**, *66*, 5377–5396. [CrossRef]
35. Ertl, P.; Schuffenhauer, A. Estimation of synthetic accessibility score of drug-like molecules based on molecular complexity and fragment contributions. *J. Cheminform.* **2009**, *1*, 8. [CrossRef]
36. Schulman, J.; Wolski, F.; Dhariwal, P.; Radford, A.; Klimov, O. Proximal policy optimization algorithms. *arXiv* **2017**. [CrossRef]
37. Tran-Nguyen, V.-K.; Jacquemard, C.; Rognan, D. LIT-PCBA: An unbiased data set for machine learning and virtual screening. *J. Chem. Inf. Model.* **2020**, *60*, 4263–4273. [CrossRef]
38. Brumshtein, B.; Greenblatt, H.M.; Butters, T.D.; Shaaltiel, Y.; Aviezer, D.; Silman, I.; Futerman, A.H.; Sussman, J.L. Crystal structures of complexes of N-butyl- and N-nonyl-deoxynojirimycin bound to acid β-glucosidase. *J. Biol. Chem.* **2007**, *282*, 29052–29058. [CrossRef]
39. Jiang, Y.; Zhang, G.; You, J.; Zhang, H.; Yao, R.; Xie, H.; Zhang, L.; Xia, Z.; Dai, M.; Wu, Y.; et al. PocketFlow is a data-and-knowledge-driven structure-based molecular generative model. *Nat. Mach. Intell.* **2024**, *6*, 326–337. [CrossRef]

molecules

MDPI

Article

Identification of Novel PPARγ Partial Agonists Based on Virtual Screening Strategy: In Silico and In Vitro Experimental Validation

Yu-E Lian [†], Mei Wang [†], Lei Ma [†], Wei Yi *, Siyan Liao, Hui Gao * and Zhi Zhou *

School of Pharmaceutical Sciences, Guangzhou Medical University, Guangzhou 511436, China
* Correspondence: yiwei@gzhmu.edu.cn (W.Y.); gaoh9@gzhmu.edu.cn (H.G.); zhouzhi@gzhmu.edu.cn (Z.Z.)
† These authors contributed equally to this work.

Abstract: Thiazolidinediones (TZDs) including rosiglitazone and pioglitazone function as peroxisome proliferator-activated receptor gamma (PPARγ) full agonists, which have been known as a class to be among the most effective drugs for the treatment of type 2 diabetes mellitus (T2DM). However, side effects of TZDs such as fluid retention and weight gain are associated with their full agonistic activities toward PPARγ induced by the AF-2 helix-involved "locked" mechanism. Thereby, this study aimed to obtain novel PPARγ partial agonists without direct interaction with the AF-2 helix. Through performing virtual screening of the Targetmol L6000 Natural Product Library and utilizing molecular dynamics (MD) simulation, as well as molecular mechanics Poisson–Boltzmann surface area (MM-PBSA) analysis, four compounds including tubuloside b, podophyllotoxone, endomorphin 1 and paliperidone were identified as potential PPARγ partial agonists. An in vitro TR-FRET competitive binding assay showed podophyllotoxone displayed the optimal binding affinity toward PPARγ among the screened compounds, exhibiting IC_{50} and ki values of 27.43 μM and 9.86 μM, respectively. Further cell-based transcription assays were conducted and demonstrated podophyllotoxone's weak agonistic activity against PPARγ compared to that of the PPARγ full agonist rosiglitazone. These results collectively demonstrated that podophyllotoxone could serve as a PPARγ partial agonist and might provide a novel candidate for the treatment of various diseases such as T2DM.

Keywords: PPARγ partial agonists; virtual screening; natural product library; TR-FRET competitive binding assay; podophyllotoxone

1. Introduction

The prevalence of type 2 diabetes (T2D), a metabolic disorder, is projected to escalate into a significant global public health concern within the next three decades. There were an estimated 529 million people living with diabetes in 2021 (96% were T2D), and that number will reach 1.31 billion by 2050 [1]. The occurrence of type 2 diabetes is attributed to reduced sensitivity of insulin-sensitive tissues towards insulin, resulting in diminished glucose uptake (GU) and subsequent hyperglycemia [2]. PPARγ has been extensively studied for its ability to mediate adipocyte differentiation in response to energy surplus and effectively regulate plasma glucose levels. Based on its pharmacological properties, PPARγ has been recognized as one of the most efficient targets for anti-diabetic drug discovery and development [3,4]. The activity of PPARγ is regulated by agonists, and the interaction between PPARγ and its ligands serves as a pivotal step in modulating PPARγ activity, with the structural characteristics of PPARγ playing a crucial role in this binding process [5].

Structurally, PPARγ comprises distinct functional domains, including an N-terminal transactivation domain containing an activation function (AF1), a highly conserved DNA-binding domain (DBD) and a C-terminal ligand-binding domain (LBD) containing a ligand-dependent transactivation function (AF2) [6]. The structure of the PPARγ ligand-binding domain (LBD) is composed of 13 α-helices and four stranded β-sheets (Figure 1). The ligand-binding pocket has three branches and has been described as a large Y- or T-shaped cavity. Branch-I is hydrophilic in nature, and is formed by helices H3, H5, H11 and H12. Branch-I is located proximal to helix H12, which forms a critical part of the activation function-2 (AF2) coregulator binding surface. In contrast, Branch-II exhibits a hydrophobic character, formed by helices H2′, H3, H6 and H7 and a β-sheet region. Branch-III consists of both hydrophobic and hydrophilic regions, surrounded by a β-sheet as well as helices H2, H3 and H5 [7]. The LBD of PPARγ has several regulatory functions [8]. Depending on the specific ligand, agonists activate PPARγ, leading to conformational changes in the ligand-binding domain and subsequently inducing the transcription of distinct target genes, thereby enhancing insulin sensitivity. Some PPARγ agonists are prescribed for the management of T2D, such as rosiglitazone and pioglitazone [9]. The synthetic agonist rosiglitazone stabilizes helices H3 and H12, which constitute the AF2 site, subsequently inducing coactivator recruitment. The binding of rosiglitazone to the ligand-binding domain (LBD) effectively inhibits CDK5-mediated phosphorylation of Ser245. Reduced Ser245 phosphorylation alters the expression of a subset of genes with regulatory functions in metabolism; for example, it increases expression of the insulin sensitizing genes adipokine and adiponectin [10]. Due to the potential side effects of PPARγ full agonists, such as weight gain, fluid retention, vascular events and bone fractures [11,12], the European Marketing Authority recommended removal of rosiglitazone from the European market in 2010 [13]. However, partial PPARγ agonists MRL-24, nTZDpa and amorfrutin 1 display similar anti-diabetic effects to rosiglitazone due to their ability to block PPARγ-Ser245 phosphorylation as effectively as rosiglitazone while only moderately inducing the expression of PPARγ target genes involved in adipocyte differentiation [14]. In addition, these partial agonists do not contact helix H12 but rather stabilize helix H3 and the β-sheet region of the binding pocket of the PPARγ LBD [15]. Consequently, partial agonists exhibit hypoglycemic effects without causing severe side effects. In recent years, inverse agonists which decrease expression of PPARγ-controlled genes and antagonists which maintain the basal transcriptional output of PPARγ have emerged as safer alternatives to full agonists [16,17]. These principles and the associated structural insights offer a novel and rational approach for the development of effective PPARγ modulators in the pursuit of anti-diabetic drug discovery [5,6,18].

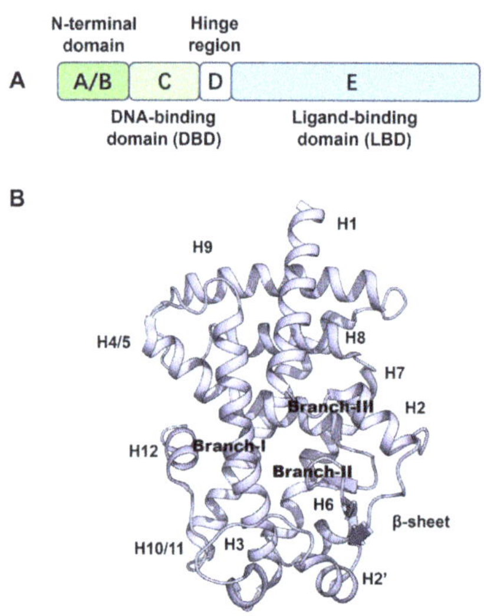

Figure 1. (**A**) Primary structure of PPARγ; (**B**) crystal structure of PPARγ (PDB code: 8DK4).

Natural compounds remain pivotal in drug discovery due to their extensive chemical diversity and potential therapeutic benefits. Historically, natural products, including plant extracts, have been significant sources of bioactive compounds [19,20]. In order to identify suitable candidates from natural products, our efforts focused on rapidly seeking new PPARγ partial agonists for anti-diabetic drug discovery. Thus, the present study's aim is (a) virtual screening of the Targetmol L6000 Natural Product Library to search for potential hits that function as PPARγ partial agonists with the ability to block cyclin-dependent kinase 5 (CDK5)-mediated phosphorylation of PPARγ-Ser245 in the absence of classical transcription activity of PPARγ via the AF-2 helix "lock" mechanism; (b) further examination of the binding stability of the system utilizing molecular dynamics and molecular mechanics Poisson–Boltzmann surface area (MM-PBSA); (c) testing the PPARγ binding affinity and agonistic activity of the selected hit. Taken together, our present study aims to combine computational approaches with in vitro experimental validation, thereby constituting a logical and robust workflow for the identification of promising candidates targeting PPARγ.

2. Results

2.1. Docking Validation, Virtual Screening and Molecular Docking

To validate the docking protocol, we redocked the partial agonist (VSP-51-2) obtained from co-crystalized PPARγ complex (PDB code: 8DK4) which yielded a score of −10.96 kcal/mol. The docking pose was basically aligned with the co-crystal ligand, proving the reliability and validity of this methodology. Virtual screening of Targetmol L6000 Natural Product Library (containing 4320 compounds) was carried out using the Autodock Vina software [21]. The docking scores and structures of the top four compounds from virtual screening are listed in Table 1. The docking scores of the four highest-binding compounds ranged from −10.68 to −10.04 kcal/mol. Compared to the original ligand VSP-51-2, the binding affinities of these compounds were generally lower, with the top binding compound endomorphin 1 producing a 0.28 kcal/mol lower binding score than VSP-51-2. Nevertheless, the affinity scores were highly negative, suggestive of favorable binding. We further evaluated the interactions between the ligands and PPARγ (Table 1). These four ligands interact with crucial structural residues through a hydrogen-bonding network with PPARγ other than those involved in classical agonism, as defined by residues shaping the activation function surface 2 (His323, Tyr473 and His449) [15]. Taken together, these four compounds exhibited favorable characteristics as candidate compounds for further exploration. In summary, based on the obtained docking scores and binding conformation, the resulting docked structures can serve as initial models for subsequent molecular dynamics (MD) simulations.

Table 1. Top four compounds from virtual screening of the Targetmol L6000 Natural Product Library via Autodock Vina.

Name	CAS	Molecular Structure	Docking Score (kcal/mol)	Hydrogen Bond	Hydrophobic Interaction
Tubuloside B	112516-04-8	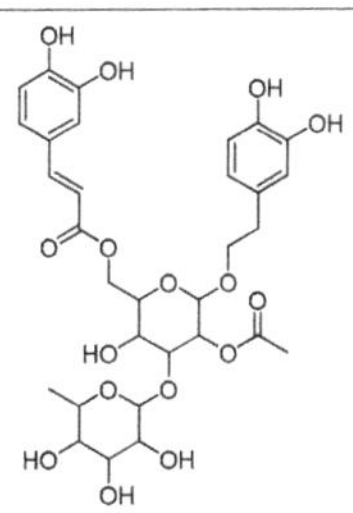	−10.07	Leu228, Gln286, Arg288, Ser342, Glu343	Phe282, Arg288, Ala292, Met329, Leu333

Table 1. *Cont.*

Name	CAS	Molecular Structure	Docking Score (kcal/mol)	Hydrogen Bond	Hydrophobic Interaction
Podophyllotoxone	477-49-6	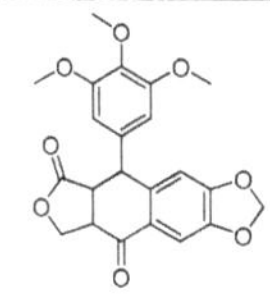	−10.04	Ser342	Arg288, Ala292, Ile326, Leu330, Ile341
Endomorphin 1	189388-22-5	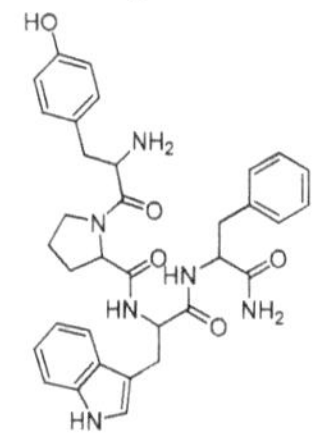	−10.68	Gly284, Cys285 Arg288	Leu228, Met329, Leu333, Arg288, Phe287, Lys263, Ile341, Cys285, Met364, Val339, Met348, Leu353, Ile326
Paliperidone	144598-75-4	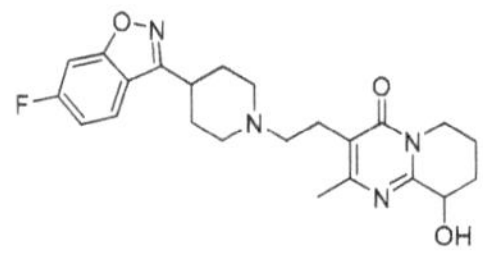	−10.44	Ser342	Leu228, Leu333, Ile341, Arg288, Ala292, Ile326, Leu330, Cys285, His449, Phe363, Tyr327

2.2. Stable Assessment of MD Simulation

The stability of these crucial binding interactions could not be fully explained by the static nature of molecular docking, given the flexibility of the residues and the corresponding fluctuations in secondary structure. To overcome these limitations, we employed MD simulations to effectively assess the stability of the protein−ligand complexes and apo protein [22]. The average values of the Root Mean Square Deviation (RMSD), Root Mean Square Fluctuation (RMSF), Radius of Gyration (RG) and Solvent-Accessible Surface Area (SASA) of the five systems in the last 50 ns of MD simulations are provided in Table S1.

RMSD evaluates the binding state and stability of the systems, with lower values indicating more stability in the system [22]. The RMSD of the protein backbone from the crystal structure of PPARγ are presented in simulation time (Figure 2). RMSD values tended to be stable after 150 ns, and the last 50 ns was selected for data analysis. Upon achieving equilibrium, the average RMSD of the backbone atoms fitted to the initial structure were 0.28, 0.24, 0.25, 0.20 and 0.24 nm for the tubuloside b-, podophyllotoxone-, endomorphin 1-, paliperidone-bound and apo PPARγ, respectively (Table S1). The apo protein exhibited a significantly higher RMSD value in comparison to the ligand-bound protein, thereby suggesting that the presence of these ligands contributes to enhancing the stability of the protein–ligand complexes. In conclusion, the overall RMSD values of the four systems ranged between 0.10 and 0.35 nm, indicating the stability of each system.

RMSF indicates the flexibility of residues in the systems. A higher RMSF value indicates a more loosely bonded structure with turns, bends and coils, whereas a lower RMSF value suggests a rigid secondary structure [23]. The RMSF values of the backbone atoms within the protein structure are depicted in Figure S2. The RMSF values of residues in four complexes fluctuated within the range of 0.05–0.60 nm, indicating overall stable dynamics. The RMSF values of the residues comprising the core structure remained consistently low across each system. The most significant differences were within the residues of Ala235–Ser245 (H2-H2′ loop) and Asp260–Lys275 (Ω loop) in PPARγ. Residues at helix H9 appeared to have the greatest fluctuation in apo protein compared with other ligand-bound systems. Overall, the average RMSF values of each system showed less deviations and suggested the overall stability.

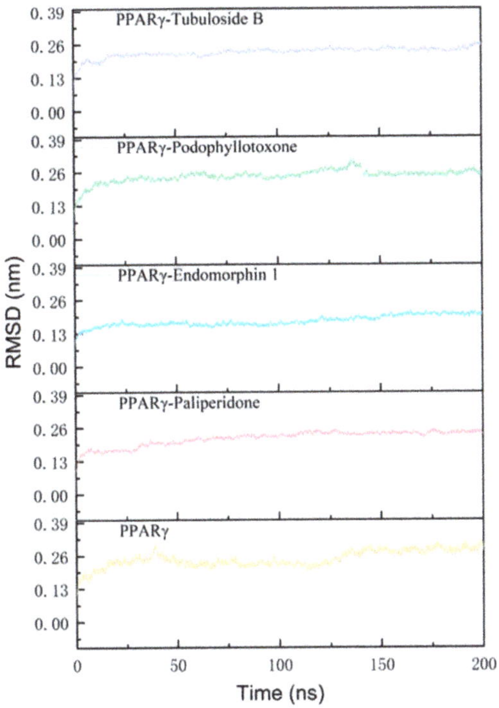

Figure 2. Analysis of RMSD of tubuloside b-, podophyllotoxone-, endomorphin 1-, paliperidone-bound and apo PPARγ.

The protein's compactness relative to its backbone was determined by calculating the RG value. It was observed that throughout the simulation, the PPARγ protein exhibited a consistently compact secondary structure (Figure S3). The apo PPARγ tended to fluctuate around its initial conformation throughout the simulation, whereas the ligand-bound PPARγ appeared to become more compact over time. SASA can be calculated to assess the packing and stability of complexes during MD simulations. The SASA analysis of the PPARγ protein was in the range of 140–160 nm^2 (Figure S4). In conclusion, both SASA and RG analyses confirmed the stability of the five systems.

2.3. *Analysis of the Binding Energy*

The calculation of binding energy (ΔE_{bind}) between the receptor and ligand is currently a commonly used and effective method for calculating binding affinity [24]. The MM-PBSA method allows a more accurate consideration of solvation contribution to the protein–ligand binding [25]. Therefore, in this study, we conducted the binding energy calculations between PPARγ and tubuloside b, podophyllotoxone, endomorphin 1 and paliperidone complexes, respectively. By comparing and analyzing the binding energy calculation results of all protein–ligand complexes (Table 2), it can be revealed that the binding energy of PPARγ with tubuloside b was the highest (−45.18 kcal/mol). This indicated that tubuloside b has higher binding affinity for PPARγ than the other three compounds, followed by endomorphin 1 (−38.86 kcal/mol), podophyllotoxone (−30.15 kcal/mol) and paliperidone (−28.08 kcal/mol). The obtained binding energy results convincingly demonstrated that these four compounds exhibit high binding affinities to PPARγ, making them suitable candidates targeting PPARγ.

Table 2. Calculated binding energies (kcal/mol) of the ligands with PPARγ by MM-PBSA using MD-derived trajectories as well as the energy decomposition.

Compound	ΔE_{vdW}	ΔE_{elec}	ΔE_{MM}	ΔG_{polar}	$\Delta G_{nonpolar}$	ΔG_{sol}	ΔE_{bind}
Tubuloside B	−69.12 ± 0.43	−79.92 ± 1.61	−149.04 ± 1.62	111.05 ± 1.25	−7.19 ± 0.02	103.86 ± 1.24	−45.18 ± 0.70
Podophyllotoxone	−53.00 ± 0.38	−21.67 ± 0.49	−74.67 ± 0.61	49.23 ± 0.41	−4.71 ± 0.01	44.52 ± 0.41	−30.15 ± 0.38
Endomorphin 1	−66.03 ± 0.38	−54.33 ± 1.39	−120.35 ± 1.49	88.34 ± 1.26	−6.58 ± 0.02	81.49 ± 1.25	−38.86 ± 0.54
Paliperidone	−55.47 ± 0.29	−18.96 ± 0.53	−74.43 ± 0.60	51.56 ± 0.53	−5.21 ± 0.01	46.35 ± 0.52	−28.08 ± 0.34

ΔE_{MM} is the MM part and amounts to $\Delta E_{vdW} + \Delta E_{elec}$. ΔG_{sol} is the solvation energies ($\Delta G_{polar} + \Delta G_{nonpolar}$). For each compound, 50 frames from the last 50 ns trajectories were used for the MM-PBSA analysis.

Regarding the energy contribution of ΔE_{bind} (Table 2), we decomposed the energy to explore the details of the interactions between binding partners. ΔE_{bind} was decomposed into four components, namely van der Waals interactions (ΔE_{vdW}), electrostatic interactions (ΔE_{elec}), polar solvation contributions (ΔG_{polar}) and nonpolar solvation contributions ($\Delta G_{nonpolar}$) [26]; see Equation (1) in the Section 3.4 for details on the decomposition. The van der Waals interaction (ΔE_{vdW}, −53.00 to −69.12 kcal/mol) and electrostatic interactions (ΔE_{elec}, −18.96 to −79.92 kcal/mol) had significant positive contributions to the final binding energy (ΔE_{bind}). Electrostatic and van der Waals interactions are often used as indicators for the evaluation of relative binding strengths [26]. In contrast, the contributions of nonpolar solvation energy ($\Delta G_{nonpolar}$, −4.71 to −7.19 kcal/mol) were relatively small, and polar solvation energy provided a negative contribution (ΔG_{polar}, 49.23 to 111.05 kcal/mol). The findings suggested that the primary factors driving the formation of the complexes between PPARγ and four compounds were ΔE_{vdW} and ΔE_{elec}.

It is noteworthy that MM-PBSA calculations yielded significantly more negative binding energies, indicating greater binding strengths than virtual screening methods. This was attributed to the fact that the Vina scoring did not utilize atomic charges for modeling electrostatic interactions [21]. Therefore, there may be a problem simulating strong electrostatic interactions between the charged parts in the Vina score. The MM-PBSA analysis provided a solution, which has been validated in multiple calculations and experiments [25]. Therefore, the employment of MD simulation and MM-PBSA significantly improve the reliability of the outcomes, rendering them necessary processes in computational research.

2.4. The Receptor and Ligand Interactions Analysis

We further explored the interaction mechanism between PPARγ and four compounds. We clustered the last 50 ns simulation trajectories and converted them into the representative binding poses for these complexes. The three-dimensional diagrams for the representative binding poses are shown in Figure S1. Biovia Discovery Studio Visualizer was used to generated two-dimensional figures to illustrate the receptor–ligand interactions (Figure 3). The complexation was primarily attributed to a diverse array of interaction types, encompassing van der Waals (vdW) contacts, hydrogen bonds, electrostatic attractions and a range of benzene-mediated interactions, specifically π-sulfur, π-alkyl and π-π stacking. Protein–ligand contact graphs are plotted for the four systems during the simulations in Figures S5–S8. Additionally, the hydrogen bond interactions played a significant role in the formation of the complex.

The analysis of hydrogen bonds serves as a crucial parameter in the detailed examination of protein–ligand complexes throughout MD simulations, providing insights into their stability and interactions. The formation of hydrogen bonds between PPARγ and specific amino acid residues has been identified as a crucial determinant in some studies [7,27]. We analyzed the number of hydrogen bond formations, as shown in Figure S9; the result found that the average number of hydrogen bonds of tubuloside b, podophyllotoxone, endomorphin 1 and paliperidone were 5, 2, 3 and 1, respectively. The results were basically consistent with the two-dimensional results (Figure 3).

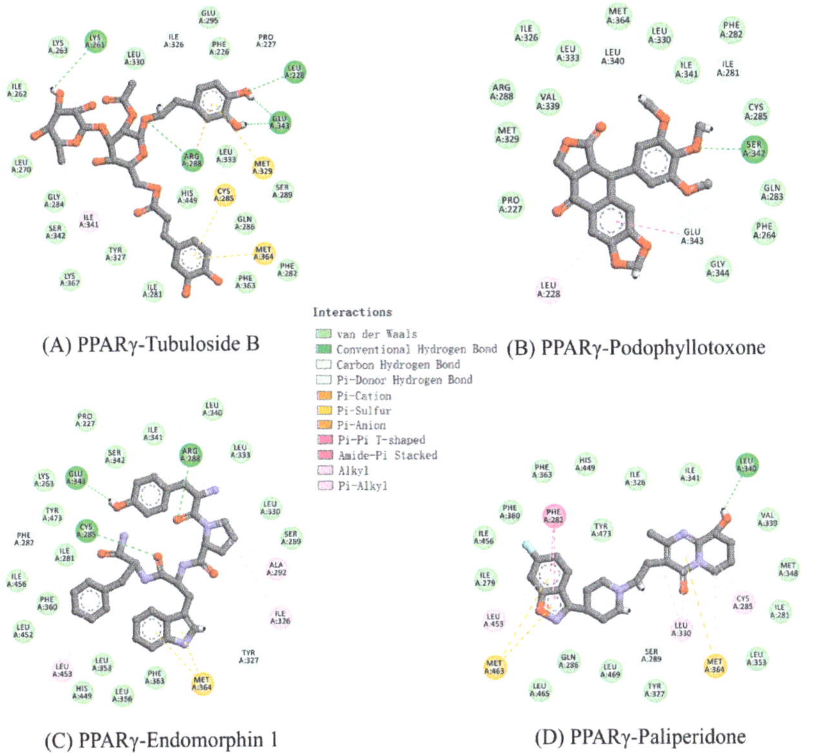

Figure 3. Two-dimensional diagrams of receptor−ligand interactions for PPARγ complexes with (**A**) tubuloside b; (**B**) podophyllotoxone; (**C**) endomorphin 1; (**D**) paliperidone. The figures were generated with Biovia Discovery Studio Visualizer, and the complexes are averaged structures clustered from the last 50 ns simulations.

In detail, one ether group in tubuloside b formed a hydrogen bond with Arg288, with a bond length of 2.2 Å. Arg288 is also a key residue located in helix H3 in PPARγ, as exhibited in partial agonists AL29-26 [28] and halofenicacid (S)-2 [29]. Two hydroxyl groups in tubuloside b formed hydrogen bonds with Leu228 and Lys261. In particular, Glu343, located in the β-sheet, formed two strong hydrogen bonds with two hydroxyl groups in tubuloside b with bond lengths of 1.47 and 1.59 Å. For podophyllotoxone, one ether group formed a hydrogen bond with a bond length of 1.88 Å with the key residue of Ser342. The role of Ser342 is an important criterion for identifying partial agonist. The binding of Ser342 with PPARγ partial agonists leads to a decrease in the stabilization of helix H12 and an increase in the stabilization of helix H3, thus affecting the recruitment of coactivators and reducing the transactivation activity of PPARγ [10,30]. One carbonyl group in endomorphin 1 exhibits a notable hydrogen bond interaction with the crucial residue of Cys285 with a bond length of 2.01 Å. The residue of Cys285 plays an important role in enhancing interactions between PPARγ and partial agonists MEKT75 [31]. Furthermore, it also formed two hydrogen bonds with Leu343 and Arg288 with bond lengths of 1.73 and 1.78 Å, respectively. The hydroxyl group of paliperidone formed a hydrogen bond with Leu340 with a bond length of 2.86 Å. It was noteworthy that the abovementioned amino acids were not the amino acids that define the compound as a full agonist rosiglitazone (His323, His449 and Tyr473) [32], suggesting the roles of compounds as potential partial agonists of PPARγ.

Following the analysis of hydrogen bonds, we examined the detailed energy contributions of each residue within the ligand-bound systems, as clearly depicted in Figures S10–S13.

It was noteworthy that the residues exhibited distinct energetic behaviors, contributing favorably (negative energy values) or unfavorably (positive energy values) upon interactions with the ligands and PPARγ. This decomposition of energy contributions provided valuable insights into the underlying molecular mechanisms that govern the interactions, facilitating a deeper understanding of the system.

Within the relatively large ligand-binding domain, residues tended to congregate predominantly around structural features such as helices H2, H3, H5 and H7 and beta sheets, where partial agonists were known to bind [10]. Notably, sulfur-containing residues, including Cys285, Met329 and Met364, exhibited particularly strong negative binding energies, indicating their crucial involvement in ligand stabilization. Further non-polar residues like Phe282, Leu330, Ile326 and Phe363 appeared to have a high energy contribution to tubuloside b, endomorphin 1 and paliperidone. The favorable binding energies exhibited by podophyllotoxone were primarily contributions from the non-polar residues of Pro227, Leu228, Leu330, Ile326, Leu333 and Ile341. This specific interaction pattern distinguished the binding mechanism of podophyllotoxone from that observed in the other three systems, highlighting a unique molecular recognition process.

By far, most partial agonists do not occupy branch I of the ligand-binding pocket, thereby lacking any contact with AF2 residues. Instead, most partial agonists predominantly occupy branches II and III portions of the ligand-binding pocket [15]. Based on the analyses of binding poses and interacting residues, tubuloside b, podophyllotoxone, endomorphin 1 and paliperidone were mainly positioned in the branch II and III portions of the ligand-binding pocket, suggesting they may function as potential partial agonists.

2.5. TR-FRET Competitive Binding Assay and PPARγ Transactivation Assay

Encouraged by the above results, we subsequently performed a TR-FRET (Time-Resolved Fluorescence Resonance Energy Transfer) competitive binding assay to detect the binding affinities of hit compounds with PPARγ at the concentrations of 100 μM and 200 μM. DMSO and PPARγ full agonist Rosi were used as negative and positive controls, respectively. As shown in Figure 4A, podophyllotoxone revealed the highest binding affinity among the four potent hits. We further evaluated the concentration of compounds that produce 50% displacement of the tracer (IC_{50}) and inhibition constant (k_i) values of podophyllotoxone. As displayed in Figure 4B, the IC_{50} and k_i values of podophyllotoxone were 27.43 μM and 9.86 μM, respectively. We further conducted cell-based transcription assays to evaluate the agonistic activity of podophyllotoxone toward PPARγ at concentrations of 30 μM and 100 μM. As shown in Figure 4C, Rosi displayed powerful agonistic activity against PPARγ; in contrast, podophyllotoxone exhibited weak PPARγ agonistic activity, similar to fenticonazole (FN), a PPARγ partial agonist reported in our previous study [33]. Taken together, the combined computational studies and in vitro evaluations including TR-FRET competitive binding assay and cell-based transcription assay identified podophyllotoxone as a ligand directly bound to PPARγ with partial agonistic activity.

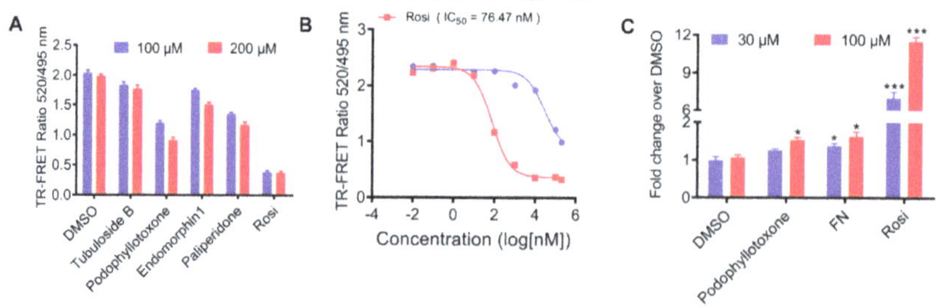

Figure 4. Identification of podophyllotoxone as a novel PPARγ ligand with potent binding affinity. (**A**) LanthaScreen TR-FRET measurements for the selected hits (100 μM and 200 μM) binding to

PPARγ after incubating for 6 h. (**B**) Dose–response competition curves for podophyllotoxone (100 μM–500 μM) incubated for 6 h. Concentrations were expressed as a log 10 scale. (**C**) Cell-based transcription assays were used to compare the agonistic activity of podophyllotoxone and FN toward PPARγ at the concentration of 30 and 100 μM. * $p < 0.05$, *** $p < 0.001$ compared with DMSO group. (n = 3, error bar = SEM).

3. Materials and Methods

3.1. Receptor and Ligand Preparations

The crystal structure of PPARγ (PDB code: 8DK4) was obtained from the RCSB Protein Data Bank database [34]. All ligand and solvent molecules in the model were removed from receptors using PyMOL (version 2.5). The missing atoms of the PPARγ protein were fixed using pdbfixer tool. The structures of compounds were collected from the PubChem compound database. The AutoDockTools module in the MGLTools package (version: 1.5.7) was used to generate the PDBQT file via adding polar hydrogens, computing Gasteiger charges and assigning AD4 atom types.

3.2. Docking Validation, Virtual Screening and Molecular Docking

A re-docking experiment docking the original ligand VSP-51-2 into the active site of PPARγ (PDB code: 8DK4) was conducted to evaluate the reliability of the docking procedure using the Autodock Vina software (version 1.2.3) [21]. Both the protein and the co-crystal ligand structure were processed with the procedure described above. The grid parameter file was set as the grid center X: −23.349, Y: −20.461, Z: 10.481 (dimensional units, Å) and dimensions 30 × 30 × 30 Å.

With this protocol, the crystal structure of PPARγ (PDB code: 8DK4) was adopted as receptor for the virtual screening of Targetmol L6000 Natural Product Library using the Autodock Vina software. Candidate compounds were considered for further examination in molecular dynamic simulations if they exhibit (a) favorable energy scores, (b) a high number of hydrogen bonds formed with PPARγ key residues and (c) the ability to bind to residues other than those involved in classical agonism, as defined by residues shaping the activation function surface 2 (His323, Tyr473 and His449).

3.3. Molecular Dynamics Simulation

To gain insights into the relative dynamics and behavioral changes of PPARγ upon interactions with different ligands, we conducted comprehensive molecular dynamic simulations encompassing five distinct systems: apo PPARγ, as well as the tubuloside b-, podophyllotoxone-, endomorphin 1- and paliperidone-bound PPARγ complexes. Molecular dynamics simulations were carried with GROMACS (version 2021.6) [35]. We optimized eight ligands at B97-3c in the water with ORCA (version 5.0.4) [36] and then calculated the restrained electrostatic potential (RESP) charges with the aid of Multiwfn (version 3.8) [37]. The Amber99SB-ILDN force field [38] was employed to model PPARγ and ions, and the general Amber force field (GAFF) [39] was chosen for the ligands. The rigid SPC model [40] was used to model water molecules. The protein–ligand complex was immersed in a cubic box and the minimum value between the solvent and the nearest box edge was 1.0 nm. The system was neutralized by adding a corresponding number of sodium or chloride (Na^+/Cl^-) ions. We performed an energy minimization step on the system using the steepest descent algorithm to simply remove any unreasonable contacts between atoms. For the equilibration phase, a short NPT equilibration was carried out for 0.5 ns. The system was equilibrated with a temperature coupling (298.15 K) using a V-rescale thermostat [41] and a pressure coupling using the C-rescale coupling algorithm [42]. Finally, 200 ns production simulations were performed under the NPT ensemble with the time step set as 2 fs. The simulated temperature was set to 298.15 K while the pressure was set to 1 bar. The Particle mesh Ewald (PME) [43] method was employed to treat the long-range electrostatic interactions and the cut off of van der Waals interaction was set to 1.2 nm. The trajectories were analyzed by tools in the GROMACS suite, including Root Mean Square Deviation (RMSD), Root Mean Square Fluctuation (RMSF), Radius of Gyration (RG) and Solvent-Accessible Surface Area (SASA).

3.4. *Calculation of the Binding Energy*

The molecular mechanics Poisson–Boltzmann surface area (MM-PBSA) is a popular method to estimate the binding energy between the receptor and ligand. It is more accurate than most fast scoring approaches in the docking and has a relatively lower computation load than the free energy calculations with an explicit solvent.

After 150 ns MD simulations, we stripped water molecules and ions from the systems, and extracted 50 conformations of complexes from the last 50 ns trajectory with an interval of 1000 ps. Decomposition of the binding energies (kcal/mol) of the ligands with PPARγ using the MM-PBSA analysis of the last 50 ns simulation trajectories. The gmx_MMPBSA toolkit [44] was used to compute the binding energy (ΔE_{bind}) that includes the contributions from van der Waals and electrostatic interactions as well as the polar ($\Delta G_{\mathrm{polar}}$) and nonpolar ($\Delta G_{\mathrm{nonpolar}}$) solvation energies. Together with an entropy contribution ($-T\Delta S$), one can obtain the binding free energy (ΔG_{bind}), as provided in Equation (1):

$$\Delta G_{\mathrm{bind}} = \Delta E_{\mathrm{MM}} + \Delta G_{\mathrm{sol}} - T\Delta S = \Delta E_{\mathrm{vdW}} + \Delta E_{\mathrm{elec}} + \Delta G_{\mathrm{polar}} + \Delta G_{\mathrm{nonpolar}} - T\Delta S \quad (1)$$

where ΔE_{MM} denotes vacuum potential energy and is the sum of van der Waals (ΔE_{vdW}) and electrostatic (ΔE_{elec}) contributions. ΔG_{sol} can be decomposed into polar ($\Delta G_{\mathrm{polar}}$) and nonpolar ($\Delta G_{\mathrm{nonpolar}}$) solvation contributions. $T\Delta S$ refers to the conformational entropy contribution at temperature T. The entropy calculations typically dominate the computational cost of the MM-PBSA estimates. Due to the large computational cost, this term is neglected in most cases of practical applications [44]. Therefore, the binding energy (ΔE_{bind}) was used for comparing different ligands against PPARγ.

3.5. *Materials*

The rosiglitazone (Rosi), fenticonazole (FN), tubuloside b, podophyllotoxone, endomorphin 1 and paliperidone were purchased from MedChem Express (Monmouth Junction, NJ, USA). The LanthaScreen™ TR-FRET PPARγ competitive binding assay kit (PV4894) and Lipofectamin 2000 was obtained from Invitrogen (Carlsbad, CA, USA). PPRE × 3 TK-luciferase plasmid, renilla luciferase plasmid and hPPARγ plasmid were constructed by Obio Technology (Shanghai, China). Dual luciferase reporter assay kits (Promega, Madison, WI, USA) were also utilized.

3.6. *TR-FRET Competitive Binding Assay*

The LanthaScreen™ TR-FRET PPARγ competitive binding assay kit (PV4894) was utilized to assess the binding affinities of tubuloside b, podophyllotoxone, endomorphin 1 and paliperidone toward PPARγ. Following the manufacturer's detailed instructions, 20 μL of each test compound, 10 μL of Fluormone™ pan-PPARγ Green and 10 μL of PPARγ LBD/Tb anti-GST Ab were added into a 384-well microtiter plate and made the final concentrations of Fluormone™ pan-PPARγ Green, PPARγ LBD and Tb anti-GST Ab 5 nM, 5 nM and 0.5 nM, respectively. Subsequently, the plate was gently mixed on an orbital plate shaker for 30 s and then incubated in the dark for 6 h at room temperature (20–25 °C). In this experimental setup, DMSO served as the negative control group, while Rosi (a known PPARγ full agonist) was used as the positive control group, allowing for the accurate evaluation of the test compounds' binding affinities relative to these standards. Subsequently, the fluorescent emission signals from each well were measured at both 495 nm and 520 nm utilizing a multi-mode reader. The TR-FRET ratio was calculated by dividing the emission signal obtained at 520 nm and 495 nm; the decreased ratio of compounds indicates their potent binding affinity. A competition curve was then generated by plotting this TR-FRET ratio against the logarithmic concentration of the test compounds. The more potent binding affinity of the compound towards PPARγ was reflected by a decrease in the TR-FRET ratio, providing a quantitative measure of their interaction strength with the receptor. The inhibition constant (K_i) for competitor was calculated by applying the Cheng–Prusoff equation as given in Equation (2):

$$K_i = IC_{50}/(1 + [tracer]/K_d) \quad (2)$$

where IC_{50} is the concentration of the competitor that produces 50% displacement of the tracer, [tracer] is the concentration of Fluormone™ pan-PPAR Green used in the assay (5 nM) and K_d is the binding constant of Fluormone™ pan-PPAR Green to PPARγ-LBD.

3.7. *PPARγ Transactivation Assay*

Based on our preceding studies [27,33,45], cos-7 cells were transfected with hPPARγ, PPRE × 3 TK-luciferase plasmids and renilla luciferase plasmids by using Lipofectamin 2000. After 24 h of incubation, the transfected cells were treated with DMSO, podophyllotoxone, FN and Rosi at a concentration of 30 or 100 μM for another 24 h. Ultimately, dual luciferase reporter assay kits were used to detect luciferase activities. Briefly, (1) 1 × passive lysis buffer was added into 24 well plates to lyse cells for 15 min; (2) 20 μL cell lysate and 100 μL luciferase assay reagent II were mixed into the 96-well plates, and the fluorescence values were immediately detected with a multifunction microplate reader (Bioteck, Synegy1), which represented the luciferase activity; (3) 100 μL Stop & Glo® Reagent was added into the above mixture and the fluorescence values were detected, which represented the renilla activity. The luciferase activity was normalized to renilla activity. Each experiment was repeated three times, with DMSO serving as the negative control and rosiglitazone (Rosi) serving as the positive control.

4. Discussion

In recent years, synthetic PPARγ partial agonists [27,33,45,46] have emerged as a promising alternative to full agonists, demonstrating hypoglycemic efficacy while mitigating the risk of severe side effects commonly associated with full agonists, including weight gain, fluid retention and heart failure. Despite the promising potential of PPARγ-targeting agents for treating T2DM, no such anti-diabetic agents specifically designed to modulate PPARγ have yet been translated into clinical practice. Upon conducting virtual screening of the Targetmol L6000 Natural Product Library, utilizing both molecular dynamics simulations and molecular mechanics Poisson–Boltzmann surface area (MM-PBSA) approaches, we have identified four compounds: tubuloside b, podophyllotoxone, endomorphin 1 and paliperidone, as promising candidates for PPARγ partial agonists.

In MM-PBSA calculations, tubuloside b showed the highest binding energy for PPARγ, followed by endomorphin 1, podophyllotoxone and paliperidone. We further utilized a TR-FRET competitive binding assay experiment to find out the binding affinity between these ligands and PPARγ. To our surprise, the binding affinity results were quite different from the MM-PBSA calculation results. Tubuloside b, endomorphin 1 and paliperidone showed low binding affinity with PPARγ, while podophyllotoxone emerged as the most potent compound among the four identified hits. Cell-based transcription assays further showed that podophyllotoxone exhibits partial agonistic activity.

The differences of binding affinities observed between molecular simulations and in vitro experiments were mainly attributed to the distinct chemical structures and orientations of the chemical groups in the compounds. In molecular dynamics simulations, such inconsistencies are often encountered and can be attributed to various factors, including the inherent limitations of the computational models and the complexity of the underlying biological systems. From a systematic evaluation of the prediction capabilities of MM-PBSA methods, researchers have concluded that the accuracy of binding free energy predictions is related to several crucial factors, including force field, charge model, continuum solvation method, interior dielectric constant and sampling method [47]. Nevertheless, computational simulations continue to play a pivotal role in drug discovery.

5. Conclusions

In this work, virtual screening of Targetmol L6000 Natural Product Library leaded to the discovery of tubuloside b, podophyllotoxone, endomorphin 1 and paliperidone as potential PPARγ partial agonists. The subsequent evaluation of RMSD, RMSF, RG and SASA values of complexes between four compounds and PPARγ LBD by performing molecular dynamic simulations indicated that the four complexes were stable. The MM-PBSA calculations revealed the binding energies of four compounds ranged from −28.08 to −45.18 kcal/mol, suggesting their potent binding affinities with PPARγ. Further in vitro evaluations including TR-FRET competitive binding assays and cell-based transcription assays demonstrated podophyllotoxone functions as a PPARγ partial agonist. Our research has yielded an effective strategy for the identification of novel PPARγ partial agonists. Additionally, we have identified a scaffold that can be utilized for structural optimization, with the goal of enhancing binding affinity while preserving the partial agonistic activity of the agonists.

Supplementary Materials: The following supporting information can be downloaded at: https://www.mdpi.com/article/10.3390/molecules29204881/s1, Table S1: Average value of RMSD, RMSF, RG and SASA; Figure S1: Three-dimensional diagrams of PPARγ protein with (A) tubuloside b; (B) podophyllotoxone; (C) endomorphin 1; (D) paliperidone; Figure S2: Analysis of RMSF of tubuloside b-, podophyllotoxone-, endomorphin 1-, paliperidone-bound and apo PPARγ; Figure S3: Analysis of RG of tubuloside b-, podophyllotoxone-, endomorphin 1-, paliperidone-bound and apo PPARγ; Figure S4: Analysis of SASA of tubuloside b-, podophyllotoxone-, endomorphin 1-, paliperidone-bound and apo PPARγ; Figure S5: Protein–ligand contacts between PPARγ and tubuloside b; H-bonds are represented as green color, purple-colored bars are depicted for hydrophobic interactions, ionic bonds are shown in blue color; Figure S6: Protein–ligand contacts between PPARγ and podophyllotoxone. H-bonds are represented as green color, purple-colored bars are depicted for hydrophobic interactions, ionic bonds are shown in blue color; Figure S7: Protein–ligand contacts between PPARγ and endomorphin 1. H-bonds are represented as green color, purple-colored bars are depicted for hydrophobic interactions, ionic bonds are shown in blue color; Figure S8: Protein–ligand contacts between PPARγ and paliperidone. H-bonds are represented as green color, purple-colored bars are depicted for hydrophobic interactions; Figure S9: Number of hydrogen bonds of protein–ligand interaction for the complexes; Figure S10: Energy contribution per residue to the binding of PPARγ with tubuloside b; Figure S11: Energy contribution per residue to the binding of PPARγ with podophyllotoxone; Figure S12: Energy contribution per residue to the binding of PPARγ with endomorphin 1; Figure S13: Energy contribution per residue to the binding of PPARγ with paliperidone.

Author Contributions: Conceptualization, Y.-E.L. and H.G.; methodology, Y.-E.L. and M.W.; software, Y.-E.L.; validation, Y.-E.L. and L.M.; formal analysis, Y.-E.L. and L.M.; investigation, Y.-E.L. and S.L.; resources, H.G.; data curation, Y.-E.L.; writing—original draft preparation, Y.-E.L.; writing—review and editing, H.G. and M.W.; visualization, Y.-E.L. and M.W.; supervision, H.G.; project administration, W.Y., H.G. and Z.Z.; funding acquisition, W.Y., H.G. and Z.Z. All authors have read and agreed to the published version of the manuscript.

Funding: This research was funded by National Natural Science Foundation of China (grant numbers 82273795, 22201051), Guangdong Basic and Applied Basic Research Foundation (grant numbers 2024A1515030179, 2024A1515010260), Science and Technology Program of Guangzhou (grant number 2023A04J0074), Plan on Enhancing Scientific Research in GMU, the College Students' Innovation Training Program (2022A106, 202310570037) and the Undergraduate Teaching Quality and Teaching Reform Project Construction Project of Guangdong Province in 2022-"Guangzhou Medical University and Xinbaotang Company" Practice Teaching Base for Integrating Science, Industry, and Education (YJGH [2023] No.4).

Institutional Review Board Statement: Not applicable.

Informed Consent Statement: Not applicable.

Data Availability Statement: Data is contained within the article and Supplementary Materials.

Conflicts of Interest: The authors declare no conflict of interest.

References

1. Ong, K.L.; Stafford, L.K.; McLaughlin, S.A.; Boyko, E.J.; Vollset, S.E.; Smith, A.E.; Dalton, B.E.; Duprey, J.; Cruz, J.A.; Hagins, H.; et al. Global, regional, and national burden of diabetes from 1990 to 2021, with projections of prevalence to 2050: A systematic analysis for the Global Burden of Disease Study 2021. *Lancet* **2023**, *402*, 203–234. [CrossRef] [PubMed]
2. Demir, S.; Nawroth, P.P.; Herzig, S.; Ekim Üstünel, B. Emerging Targets in Type 2 Diabetes and Diabetic Complications. *Adv. Sci.* **2021**, *8*, 2100275. [CrossRef] [PubMed]
3. Rangwala, S.M.; Lazar, M.A. Peroxisome proliferator-activated receptor γ in diabetes and metabolism. *Trends Pharmacol. Sci.* **2004**, *25*, 331–336. [CrossRef]
4. Janani, C.; Ranjitha Kumari, B.D. PPAR gamma gene—A review. *Diabetes Metab. Syndr. Clin. Res. Rev.* **2015**, *9*, 46–50. [CrossRef]
5. Chigurupati, S.; Dhanaraj, S.A.; Balakumar, P. A step ahead of PPARγ full agonists to PPARγ partial agonists: Therapeutic perspectives in the management of diabetic insulin resistance. *Eur. J. Pharmacol.* **2015**, *755*, 50–57. [CrossRef]
6. Ahmadian, M.; Suh, J.M.; Hah, N.; Liddle, C.; Atkins, A.R.; Downes, M.; Evans, R.M. PPARγ signaling and metabolism: The good, the bad and the future. *Nat. Med.* **2013**, *19*, 557–566. [CrossRef]
7. Kroker, A.J.; Bruning, J.B. Review of the Structural and Dynamic Mechanisms of PPARγ Partial Agonism. *PPAR Res.* **2015**, *2015*, 816856. [CrossRef]
8. Chan, L.S.A.; Wells, R.A.; Shi, X.-M. Cross-Talk between PPARs and the Partners of RXR: A Molecular Perspective. *PPAR Res.* **2009**, *2009*, 925309. [CrossRef]
9. Lebovitz, H.E. Thiazolidinediones: The Forgotten Diabetes Medications. *Curr. Diabetes Rep.* **2019**, *19*, 151. [CrossRef]
10. Choi, J.H.; Banks, A.S.; Estall, J.L.; Kajimura, S.; Boström, P.; Laznik, D.; Ruas, J.L.; Chalmers, M.J.; Kamenecka, T.M.; Blüher, M.; et al. Anti-diabetic drugs inhibit obesity-linked phosphorylation of PPARγ by Cdk5. *Nature* **2010**, *466*, 451–456. [CrossRef]
11. Rennings, A.J.M.; Smits, P.; Stewart, M.W.; Tack, C.J. Fluid Retention and Vascular Effects of Rosiglitazone in Obese, Insulin-Resistant, Nondiabetic Subjects. *Diabetes Care* **2006**, *29*, 581–587. [CrossRef]
12. Guan, Y.; Hao, C.; Cha, D.R.; Rao, R.; Lu, W.; Kohan, D.E.; Magnuson, M.A.; Redha, R.; Zhang, Y.; Breyer, M.D. Thiazolidinediones expand body fluid volume through PPARγ stimulation of ENaC-mediated renal salt absorption. *Nat. Med.* **2005**, *11*, 861–866. [CrossRef]
13. Rosen, C.J. Revisiting the Rosiglitazone Story—Lessons Learned. *N. Engl. J. Med.* **2010**, *363*, 803–806. [CrossRef]
14. Weidner, C.; de Groot, J.C.; Prasad, A.; Freiwald, A.; Quedenau, C.; Kliem, M.; Witzke, A.; Kodelja, V.; Han, C.-T.; Giegold, S.; et al. Amorfrutins are potent antidiabetic dietary natural products. *Proc. Natl. Acad. Sci. USA* **2012**, *109*, 7257–7262. [CrossRef]
15. Bruning, J.B.; Chalmers, M.J.; Prasad, S.; Busby, S.A.; Kamenecka, T.M.; He, Y.; Nettles, K.W.; Griffin, P.R. Partial Agonists Activate PPARγ Using a Helix 12 Independent Mechanism. *Structure* **2007**, *15*, 1258–1271. [CrossRef]
16. Frkic, R.L.; Pederick, J.L.; Horsfall, A.J.; Jovcevski, B.; Crame, E.E.; Kowalczyk, W.; Pukala, T.L.; Chang, M.R.; Zheng, J.; Blayo, A.-L.; et al. PPARγ Corepression Involves Alternate Ligand Conformation and Inflation of H12 Ensembles. *ACS Chem. Biol.* **2023**, *18*, 1115–1123. [CrossRef]
17. Frkic, R.L.; Marshall, A.C.; Blayo, A.-L.; Pukala, T.L.; Kamenecka, T.M.; Griffin, P.R.; Bruning, J.B. PPARγ in Complex with an Antagonist and Inverse Agonist: A Tumble and Trap Mechanism of the Activation Helix. *iScience* **2018**, *5*, 69–79. [CrossRef]
18. Chen, F.; Ma, L.; Cai, G.; Tang, J.; Wang, Y.; Liu, Q.; Liu, X.; Hou, N.; Zhou, Z.; Yi, W. Identification of a novel PPARγ modulator with good anti-diabetic therapeutic index via structure-based screening, optimization and biological validation. *Biomed. Pharmacother.* **2022**, *154*, 113653. [CrossRef]
19. Omoboyowa, D.A.; Singh, G.; Fatoki, J.O.; Oyeneyin, O.E. Computational investigation of phytochemicals from Abrus precatorius seeds as modulators of peroxisome proliferator-activated receptor gamma (PPARγ). *J. Biomol. Struct. Dyn.* **2022**, *41*, 5568–5582. [CrossRef]
20. Chu, Y.; Gui, S.; Zheng, Y.; Zhao, J.; Zhao, Y.; Li, Y.; Chen, X. The natural compounds, Magnolol or Honokiol, promote adipose tissue browning and resist obesity through modulating PPARα/γ activity. *Eur. J. Pharmacol.* **2024**, *969*, 176438. [CrossRef]
21. Eberhardt, J.; Santos-Martins, D.; Tillack, A.F.; Forli, S. AutoDock Vina 1.2.0: New Docking Methods, Expanded Force Field, and Python Bindings. *J. Chem. Inf. Model.* **2021**, *61*, 3891–3898. [CrossRef] [PubMed]
22. Mandal, S.; Faizan, S.; Raghavendra, N.M.; Kumar, B.R.P. Molecular dynamics articulated multilevel virtual screening protocol to discover novel dual PPAR α/γ agonists for anti-diabetic and metabolic applications. *Mol. Divers.* **2022**, *27*, 2605–2631. [CrossRef] [PubMed]
23. Pathak, R.K.; Gupta, A.; Shukla, R.; Baunthiyal, M. Identification of new drug-like compounds from millets as Xanthine oxidoreductase inhibitors for treatment of Hyperuricemia: A molecular docking and simulation study. *Comput. Biol. Chem.* **2018**, *76*, 32–41. [CrossRef] [PubMed]
24. Zhou, B.; Zhang, Y.; Jiang, W.; Zhang, H. Virtual Screening of FDA-Approved Drugs for Enhanced Binding with Mitochondrial Aldehyde Dehydrogenase. *Molecules* **2022**, *27*, 8773. [CrossRef]
25. Genheden, S.; Ryde, U. The MM/PBSA and MM/GBSA methods to estimate ligand-binding affinities. *Expert Opin. Drug Discov.* **2015**, *10*, 449–461. [CrossRef] [PubMed]
26. Jiang, W.; Chen, J.; Zhang, P.; Zheng, N.; Ma, L.; Zhang, Y.; Zhang, H. Repurposing Drugs for Inhibition against ALDH2 via a 2D/3D Ligand-Based Similarity Search and Molecular Simulation. *Molecules* **2023**, *28*, 7325. [CrossRef] [PubMed]

27. Ma, L.; Tang, J.; Cai, G.; Chen, F.; Liu, Q.; Zhou, Z.; Zhang, S.; Liu, X.; Hou, N.; Yi, W. Structure-based screening and biological validation of the anti-thrombotic drug-dicoumarol as a novel and potent PPARγ-modulating ligand. *Bioorganic Chem.* **2022**, *129*, 106191. [CrossRef]
28. Capelli, D.; Cerchia, C.; Montanari, R.; Loiodice, F.; Tortorella, P.; Laghezza, A.; Cervoni, L.; Pochetti, G.; Lavecchia, A. Structural basis for PPAR partial or full activation revealed by a novel ligand binding mode. *Sci. Rep.* **2016**, *6*, 34792. [CrossRef]
29. Laghezza, A.; Montanari, R.; Lavecchia, A.; Piemontese, L.; Pochetti, G.; Iacobazzi, V.; Infantino, V.; Capelli, D.; De Bellis, M.; Liantonio, A.; et al. On the Metabolically Active Form of Metaglidasen: Improved Synthesis and Investigation of Its Peculiar Activity on Peroxisome Proliferator-Activated Receptors and Skeletal Muscles. *ChemMedChem* **2015**, *10*, 555–565. [CrossRef]
30. de Groot, J.C.; Weidner, C.; Krausze, J.; Kawamoto, K.; Schroeder, F.C.; Sauer, S.; Büssow, K. Structural Characterization of Amorfrutins Bound to the Peroxisome Proliferator-Activated Receptor γ. *J. Med. Chem.* **2013**, *56*, 1535–1543. [CrossRef] [PubMed]
31. Ohashi, M.; Gamo, K.; Oyama, T.; Miyachi, H. Peroxisome proliferator-activated receptor gamma (PPARγ) has multiple binding points that accommodate ligands in various conformations: Structurally similar PPARγ partial agonists bind to PPARγ LBD in different conformations. *Bioorganic Med. Chem. Lett.* **2015**, *25*, 2758–2762. [CrossRef] [PubMed]
32. Ahsan, W. The Journey of Thiazolidinediones as Modulators of PPARs for the Management of Diabetes: A Current Perspective. *Curr. Pharm. Des.* **2019**, *25*, 2540–2554. [CrossRef] [PubMed]
33. Ma, L.; Lian, Y.; Tang, J.; Chen, F.; Gao, H.; Zhou, Z.; Hou, N.; Yi, W. Identification of the anti-fungal drug fenticonazole nitrate as a novel PPARγ-modulating ligand with good therapeutic index: Structure-based screening and biological validation. *Pharmacol. Res.* **2021**, *173*, 105860. [CrossRef]
34. Burley, S.K.; Bhikadiya, C.; Bi, C.; Bittrich, S.; Chao, H.; Chen, L.; Craig, P.A.; Crichlow, G.V.; Dalenberg, K.; Duarte, J.M.; et al. RCSB Protein Data Bank (RCSB.org): Delivery of experimentally-determined PDB structures alongside one million computed structure models of proteins from artificial intelligence/machine learning. *Nucleic Acids Res.* **2023**, *51*, 488–508. [CrossRef]
35. Páll, S.; Zhmurov, A.; Bauer, P.; Abraham, M.; Lundborg, M.; Gray, A.; Hess, B.; Lindahl, E. Heterogeneous parallelization and acceleration of molecular dynamics simulations in GROMACS. *J. Chem. Phys.* **2020**, *153*, 134110. [CrossRef]
36. Neese, F. Software update: The ORCA program system—Version 5.0. *Wiley Interdiscip. Rev. Comput. Mol. Sci.* **2022**, *12*, e1606. [CrossRef]
37. Lu, T.; Chen, F. Multiwfn: A multifunctional wavefunction analyzer. *J. Comput. Chem.* **2011**, *33*, 580–592. [CrossRef]
38. Lindorff-Larsen, K.; Piana, S.; Palmo, K.; Maragakis, P.; Klepeis, J.L.; Dror, R.O.; Shaw, D.E. Improved side-chain torsion potentials for the Amber ff99SB protein force field. *Proteins* **2010**, *78*, 1950–1958. [CrossRef]
39. Wang, J.; Wolf, R.M.; Caldwell, J.W.; Kollman, P.A.; Case, D.A. Development and testing of a general amber force field. *J. Comput. Chem.* **2004**, *25*, 1157–1174. [CrossRef]
40. Jorgensen, W.L.; Chandrasekhar, J.; Madura, J.D.; Impey, R.W.; Klein, M.L. Comparison of simple potential functions for simulating liquid water. *J. Chem. Phys.* **1983**, *79*, 926–935. [CrossRef]
41. Bussi, G.; Donadio, D.; Parrinello, M. Canonical sampling through velocity rescaling. *J. Chem. Phys.* **2007**, *126*, 014101. [CrossRef] [PubMed]
42. Bernetti, M.; Bussi, G. Pressure control using stochastic cell rescaling. *J. Chem. Phys.* **2020**, *153*, 114107. [CrossRef]
43. Essmann, U.; Perera, L.; Berkowitz, M.L.; Darden, T.; Lee, H.; Pedersen, L.G. A smooth particle mesh Ewald method. *J. Chem. Phys.* **1995**, *103*, 8577–8593. [CrossRef]
44. Valdés-Tresanco, M.S.; Valdés-Tresanco, M.E.; Valiente, P.A.; Moreno, E. gmx_MMPBSA: A New Tool to Perform End-State Free Energy Calculations with GROMACS. *J. Chem. Theory Comput.* **2021**, *17*, 6281–6291. [CrossRef] [PubMed]
45. Ma, L.; Tang, J.; Chen, F.; Liu, Q.; Huang, J.; Liu, X.; Zhou, Z.; Yi, W. Structure-based screening, optimization and biological evaluation of novel chrysin-based derivatives as selective PPARγ modulators for the treatment of T2DM and hepatic steatosis. *Eur. J. Med. Chem.* **2024**, *276*, 116728. [CrossRef] [PubMed]
46. Jiang, H.; Zhou, X.E.; Shi, J.; Zhou, Z.; Zhao, G.; Zhang, X.; Sun, Y.; Suino-Powell, K.; Ma, L.; Gao, H.; et al. Identification and structural insight of an effective PPARγ modulator with improved therapeutic index for anti-diabetic drug discovery. *Chem. Sci.* **2020**, *11*, 2260–2268. [CrossRef]
47. Wang, E.; Sun, H.; Wang, J.; Wang, Z.; Liu, H.; Zhang, J.Z.H.; Hou, T. End-Point Binding Free Energy Calculation with MM/PBSA and MM/GBSA: Strategies and Applications in Drug Design. *Chem. Rev.* **2019**, *119*, 9478–9508. [CrossRef]

molecules

MDPI

Article

Structure-Based Virtual Screening for Methyltransferase Inhibitors of SARS-CoV-2 nsp14 and nsp16

Kejue Wu [1,†], Yinfeng Guo [1,†], Tiefeng Xu [2], Weifeng Huang [1], Deyin Guo [2,3], Liu Cao [2] and Jinping Lei [1,*]

1 Guangdong Key Laboratory of Chiral Molecule and Drug Discovery, School of Pharmaceutical Sciences, Sun Yat-Sen University, Guangzhou 510006, China; wukj6@mail2.sysu.edu.cn (K.W.); guoyf36@163.com (Y.G.); huangwf26@mail2.sysu.edu.cn (W.H.)

2 Centre for Infection and Immunity Studies (CIIS), School of Medicine, Sun Yat-Sen University, Shenzhen 518107, China; guo_deyin@gzlab.ac.cn (D.G.); caoliu@mail.sysu.edu.cn (L.C.)

3 Guangzhou Laboratory, Bio-Island, Guangzhou 510320, China

* Correspondence: leijp@mail.sysu.edu.cn

† These authors contributed equally to this work.

Abstract: The ongoing COVID-19 pandemic still threatens human health around the world. The methyltransferases (MTases) of SARS-CoV-2, specifically nsp14 and nsp16, play crucial roles in the methylation of the N7 and 2′-O positions of viral RNA, making them promising targets for the development of antiviral drugs. In this work, we performed structure-based virtual screening for nsp14 and nsp16 using the screening workflow (HTVS, SP, XP) of Schrödinger 2019 software, and we carried out biochemical assays and molecular dynamics simulation for the identification of potential MTase inhibitors. For nsp14, we screened 239,000 molecules, leading to the identification of three hits A1–A3 showing N7-MTase inhibition rates greater than 60% under a concentration of 50 μM. For the SAM binding and nsp10-16 interface sites of nsp16, the screening of 210,000 and 237,000 molecules, respectively, from ZINC15 led to the discovery of three hit compounds B1–B3 exhibiting more than 45% of 2′-O-MTase inhibition under 50 μM. These six compounds with moderate MTase inhibitory activities could be used as novel candidates for the further development of anti-SARS-CoV-2 drugs.

Keywords: SARS-CoV-2; nsp14; nsp16; MTase inhibitors; structure-based virtual screening

1. Introduction

According to the latest statistics of the COVID-19 pandemic (https://covid19.who.int/, accessed on 25 February 2024), there are more than 774 million people globally suffering from infections of severe acute respiratory syndrome coronavirus 2 (SARS-CoV-2), which has caused about 7.01 million deaths as of 25 February 2024 [1]. Compared to the earlier Middle East respiratory syndrome coronavirus (MERS-CoV) and severe acute respiratory syndrome coronavirus 1 (SARS-CoV-1) pandemics, the current COVID-19 pandemic poses a substantial threat to human health [2,3]. As SARS-CoV-2 continues to spread and undergo mutations to a variety of variants, such as Delta and Omicron, the demand for effective vaccines as well as antiviral therapeutic drugs increases drastically [4,5]. However, there is no specific drug to cure SARS-CoV-2 infection currently, and only a few antiviral agents such as Remdesivir, Azvudine, Paxlovid, and Molnupiravir have been approved for the treatment of COVID-19 in adult and pediatric patients [6–9].

SARS-CoV-2 is a positive-strain RNA virus with a genomic size of approximately 29.9 kb that is extremely contagious [10]. The first two-thirds of the SARS-CoV-2 genome is composed of a pair of large open reading frames (Orf1a and Orf1ab), whose encoded polyproteins (pp1a and pp1ab) are decomposed into 16 different non-structural proteins (nsp1 to nsp16) that are responsible for viral replication and transcription in eukaryotic cells [11]. Among these nsp proteins, nsp12 (RNA guanylyltransferase), nsp13 (RNA triphosphatase), nsp14 (RNA guanine-N7-methyltransferase, N7-MTase), and nsp16 (RNA

2′-O-methyltransferase, 2′-O-MTase) are involved in viral mRNA capping (Figure 1), which helps the SARS-CoV-2 virus escape the administration of the host innate immune system.

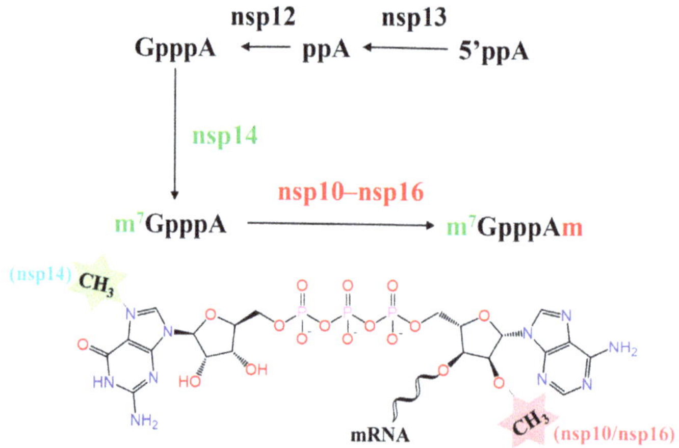

Figure 1. SARS-CoV-2 capping outline. The initial cap core structure (cap-0) of SARS-CoV-2 is formed at the 5′-end of RNA. First, the newly generated RNA is hydrolyzed into ppRNA by RNA 5′-triphosphatase (RTPase/nsp13), and the terminal γ-phosphate is removed. Then, Gppp-RNA is formed under the catalysis of guanylyltransferase (GTase/nsp12). Subsequently, the N7-position is methylated by N7-MTase (nsp14) to form the cap-0 structure. Finally, nsp10–16 catalyzes the formation of the ultimate cap-1 structure.

nsp14 is an N7-MTase that converts SARS-CoV-2 viral mRNA to a cap-0 structure, while the 2′-O-MTase nsp16 along with co-factor nsp10 is essential for cap-1 structure formation (Figure 1) [12]. Both N7-MTase and 2′-O-MTase use S-adenosyl-l-methionine (SAM) as the methyl donor to methylate the SARS-CoV-2 viral mRNA at the SAM binding site [13]. Stable monomeric protein nsp10 interacts with nsp16 to extend its RNA binding groove and stabilize its SAM binding pocket, both of which are essential for nsp16 MTase activity [14–16]. Therefore, the SAM binding sites of nsp14 and nsp16 and the nsp10-nsp16 interface are potential targets for developing highly specific anti-COVID-19 drugs [17]. Several small molecular inhibitors targeting nsp14 or nsp16 have been reported and validated by in vitro experiments, such as Sinefungin, SS148, and WZ16 [14,18,19]. However, most of them are SAM analogs that possess similar scaffolds [20,21]. Thus, further research is still warranted to discover inhibitors with diverse scaffolds and structures.

In this study, we conducted structure-based virtual screening (SBVS) to identify small molecular inhibitors targeting the nsp14 or nsp16 of SARS-CoV-2. A total of 349,000 compounds from the ZINC15 database and 100,000 compounds from the ChemDiv database were collected and screened by filtering steps of the virtual screening workflow (HTVS, SP, XP) of Schrödinger software. A total of 9 and 8 compounds were screened out for the further in vitro experimental validation of N7-MTase and 2′-O-MTase inhibition activities, respectively. Finally, 3 compounds A1–A3 exhibited more than 60% of inhibition against N7-MTase, and 3 compounds B1–B3 exhibited more than 45% of inhibition against 2′-O-MTase. These compounds could be used as potential MTase inhibitors for the future drug design of SARS-CoV-2.

2. Results

The overall workflow of SBVS for identifying potential SARS-CoV-2 nsp14 or nsp16 inhibitors is presented in Figure 2. In this SBVS process, we first pre-processed the SAM binding sites of nsp14 and nsp16 by analyzing the binding modes of reported SARS-CoV-2 MTase inhibitors, and we predicted the potential binding pocket at the nsp10–nsp16 interface by utilizing the Protein Plus DoGSiteScorer webserver (https://proteins.plus/, accessed on 6 January 2023) [22]. Subsequently, we performed the SBVS successively using Glide HTVS, SP, XP docking, and visual inspection of the binding interactions. Finally, we conducted in vitro assays to validate the MTase inhibition activities of the selected compounds.

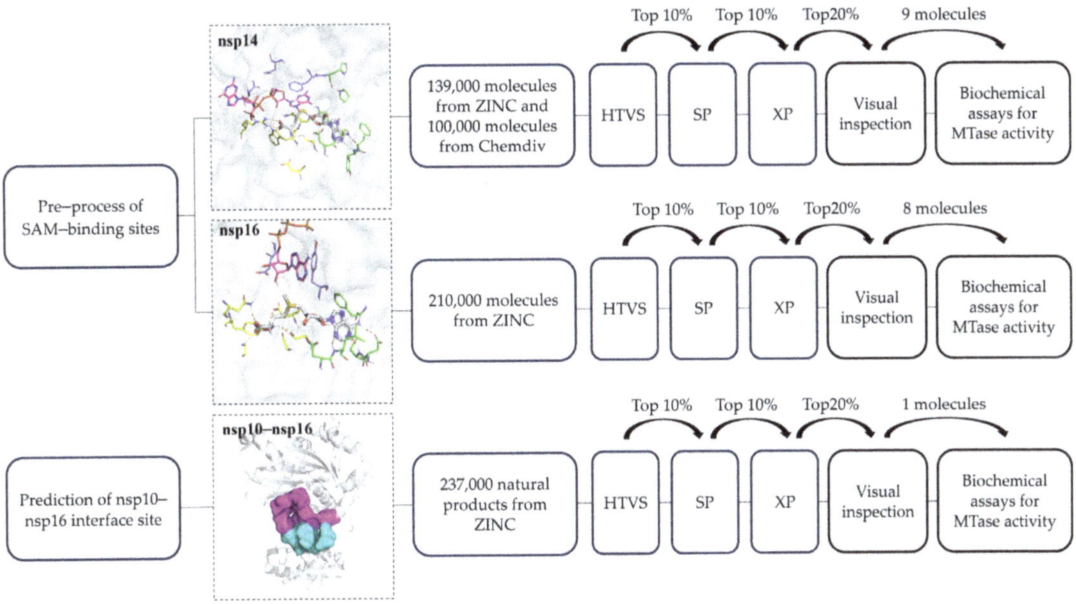

Figure 2. Framework of structure-based virtual screening.

2.1. Binding Site Processing of SARS-CoV-2 nsp14 and nsp16

The X-ray data analysis of the recently reported crystal structures of SARS-CoV-2 nsp14 and nsp16 was compared and is presented in Tables S1 and S2, and the qualities of these structures were evaluated using the MolProbity webserver (http://molprobity.biochem.duke.edu/index.php, accessed on 20 October 2022) [23] to select a reasonable structure for subsequent SBVS. Consequently, the structures coded with 7R2V and 6WVN [24,25] were selected for nsp14 and nsp16, respectively, based on the rank of structural resolution, Clashscore and MolProbity score, and considering the importance of co-factor SAM/SAH and the substrate for N7 and 2′-O methylation.

In the SAM binding site of nsp14, the SAM/SAH binding pocket is in close proximity to the RNA binding pocket, and the reported inhibitors are capable of occupying both the SAM and RNA cap binding pockets [26–31]. Thus, we pre-processed the nsp14 SAM binding site by dividing it into three parts (Figure 3a): (i) SAM-adenine binding cavity, which includes residues Asp352, Ala353, and Tyr368 that can form hydrogen bond interactions with the adenosine group of the reported SAM analog inhibitors; (ii) SAM-tail binding cavity, containing residues Arg310, Gly333, and Trp385/Asn386 that can form hydrogen bonds with the methionine part of SAM analogs; and (iii) RNA cap binding cavity, including Phe426 that forms a π-π stacking interaction with the base group of the RNA cap. In the subsequent virtual screening, we would select the potential inhibitors that occupy both the SAM-adenine and SAM-tail binding cavities or both the SAM-adenine and RNA cap binding cavities.

The SAM binding site of nsp16 presented a restricted spatial configuration compared with that of nsp14 [17,18,20,32–34] because of the clashes between the SAM and RNA cap binding pockets in the presence of m7GpppA in nsp16, leading to an open state of the cap-0 binding pocket [24]. Hence, we divided the nsp16 SAM binding site only into two distinct parts (Figure 3b): (i) SAM-adenine binding cavity, which includes residues Leu6898, Asp6912, and Cys6913 that can form hydrogen bonds with the adenosine group of the reported SAM analog inhibitors; and (ii) SAM-tail binding cavity, containing residues Gly6869, Gly6879, Asn6841, and Asp6928 that can form H-bond interactions with the methionine part of SAM analogs. In the upcoming virtual screenings, we would screen out the potential inhibitors capable of concurrently binding to both the SAM-adenine and SAM-tail binding cavities.

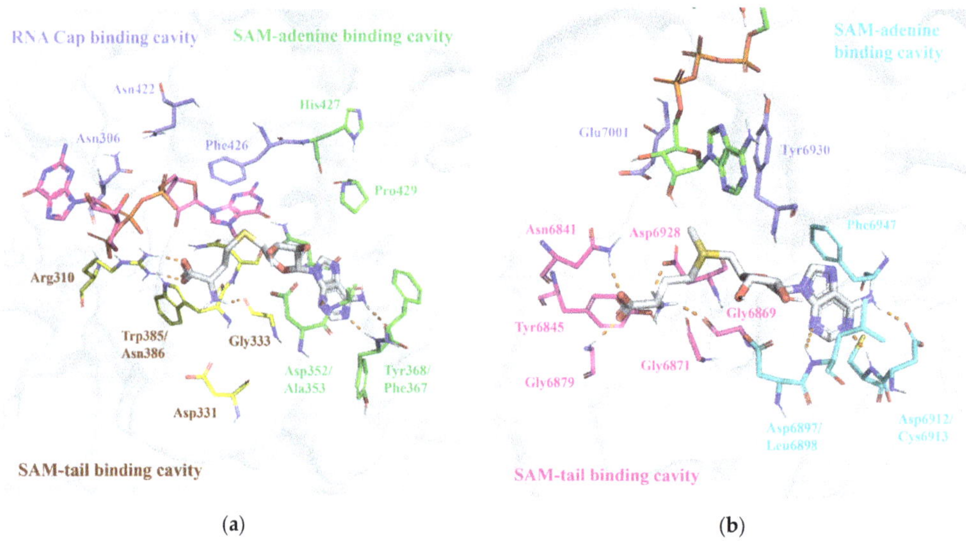

Figure 3. (**a**) Protein surface of SARS-CoV-2 nsp14 SAM binding site (PDB ID: 7R2V). SAH is denoted by white sticks; m7GpppA (align from PDB ID: 7QIF) is denoted by pink sticks; the amino acids of the SAM-adenine, SAM-tail, and RNA cap binding cavity are denoted by green, yellow, and purple sticks, respectively; and hydrogen bonds are denoted by orange dashed lines. (**b**) Protein surface of SARS-CoV-2 nsp16 SAM binding site (PDB ID: 6WVN). SAM is denoted by white sticks; m7GpppA is denoted by green sticks; the amino acids of the SAM-adenine and SAM-tail binding cavity are denoted by cyan and pink sticks, respectively; and hydrogen bonds are denoted by orange dashed lines.

The nsp10–nsp16 interface was identified as a potential target for developing 2′-O-MTase inhibitors through disrupting the nsp10–nsp16 interactions [34]. The potential binding pocket at the nsp10–nsp16 interface was detected using the Protein Plus DoGSiteScorer webserver (https://proteins.plus/, accessed on 6 January 2023) [22,35]. The cavity shown in Figure S1 presented the best region with the highest druggability score of 0.81 and a volume of 589.38 Å^3 for drug binding. The amino acid residues positioned in the predicted binding pocket are listed in Table S3 and plotted by sticks in Figure S1 using PyMOL (DeLano Scientific, Palo Alto, CA, USA). As can be seen from Figure S1, the amino acid residues in the DoGSiteScorer-predicted binding pocket were consistent with the interacting interface residues (Figure S1) extracted by PDBsum prot-prot analysis [24]. The key residues in the predicted binding pocket were Arg6884, Gln6885, Met7045 of nsp16, and Leu4298, Thr4300, Pro4312, Gly4347, and Tyr4349 of nsp10 (Table S3).

2.2. Structure-Based Virtual Screening

For SBVS by glide docking, the receptor grid in the SAM binding site of nsp14 and nsp16 was, respectively, defined as a 30 Å box centered on the O-atom of residue Asn386 and a 35 Å box centered on the 2′-O atom of the m7GpppA substrate. And the receptor grid for SBVS at the nsp10–nsp16 interface was defined as a 30 Å box centered on the O atom of residue Gln6885 in nsp16. The filtering steps of the SBVS workflow (HTVS, SP, XP) in Schrödinger were employed, and visual inspection analysis was also carried out to screen out the potential N7 and 2′-O MTase inhibitors. We first selected the top 10% hits from Glide HTVS for subsequent SP filtering, and then chose the top 10% hits from SP for subsequent XP filtering, and finally the top 20% hits from XP were subjected to visual inspection screening [36–38]. In the visual inspection screening step, we used the binding mode and interaction of Sinefungin as the positive control (Figure S2) to screen out compounds with similar or more favorable binding interactions.

2.2.1. SBVS Results of nsp14 Inhibitors

We identified nine potential nsp14 inhibitors A1–A9 (Y207-3841, ZINC000009481760, D306-0032, ZINC000257219502, ZINC000012154664, C226-1222, ZINC000257316872, D665-0380, ZINC000008892924) through multiple rounds of screening (Figures 4, 5, S3 and S7a and Table 1). As shown in Table 1, the molecular weight and logP of these nine compounds ranged from 350 to 480 and 0.3 to 3.8, respectively. Almost all molecules exhibited docking scores less than −8.50 kcal/mol and formed hydrogen bond interactions with Tyr368 in the SAM-adenine binding cavity. Among these nine compounds, seven molecules (A1, A4–A9) occupy both the SAM-adenine and RNA cap binding cavities and engage in π-π interactions with Phe426 in the RNA cap binding cavity. The remaining two molecules (A2–A3) occupy both the SAM-adenine and SAM-tail binding cavities and exhibit a similar binding conformation as SAM and Sinefungin (Figure S2a) that form hydrogen bonds with Arg310 in the SAM-tail cavity. For the top three compounds A1–A3 with high docking scores less than −9.17 kcal/mol (Figure 5, Table 1), the two N-atoms of indazole in compound A1 form hydrogen bonds with Tyr368 in the SAM-adenine cavity and benzene forms a π-π stacking interaction with Phe426 in the RNA cap binding cavity to achieve dual substrate occupancy. The two N-atoms of adenine in compound A2 form two hydrogen bonds with Tyr368, a π-π stacking interaction of adenine with Phe367 in the SAM-adenine cavity, and hydrogen bond interactions of the sulfonyl group with Arg310 and Asn386 in the SAM-tail cavity. The two N-atoms of pyrimidine in compound A3 form hydrogen bonds with Ala353 and Tyr368 in the SAM-adenine cavity and the O-atoms of the ester group form hydrogen bonds with Arg310 and Asn388 in the SAM-tail cavity. Based on the molecular property, docking score, and binding interaction analysis, these nine compounds were selected for further in vitro validations of N7-MTase inhibition activities.

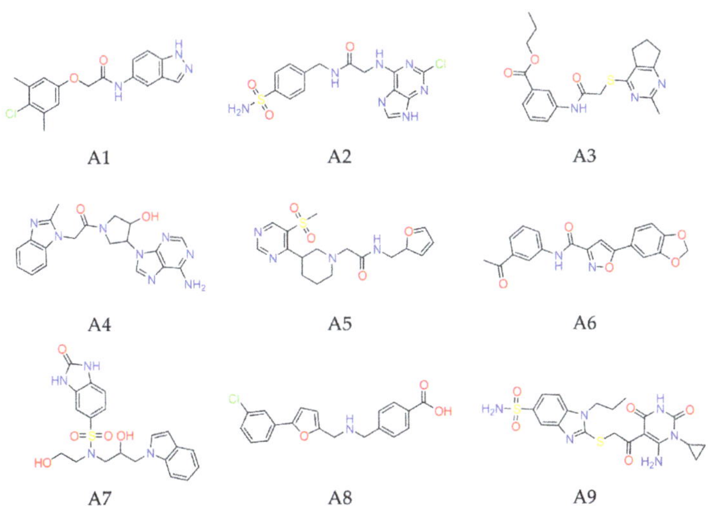

Figure 4. Chemical structures of potential inhibitors A1–A9 targeting nsp14 SAM binding site.

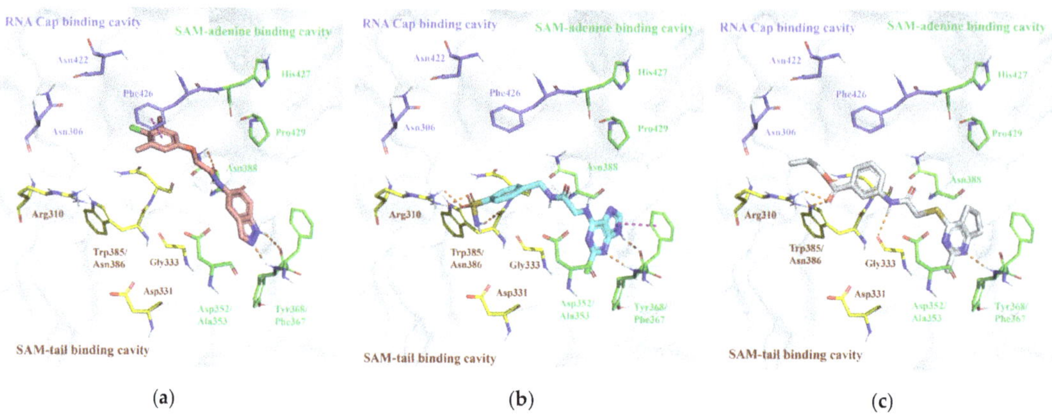

Figure 5. A three-dimensional view of the binding modes of (**a**) A1, (**b**) A2, and (**c**) A3 in the nsp14 SAM binding site. Amino acids in the SAM-adenine, SAM-tail, and RNA Cap binding cavities are denoted by green, yellow, and purple sticks, respectively. The hydrogen bonds are denoted by orange dashed lines, and π-π stacking interactions are denoted by pink dashed lines.

Table 1. The SBVS and in vitro validation results of potential inhibitors targeting nsp14 SAM binding site.

Code	Compound	Molecular Weight	LogP	Docking Score (kcal/mol)	H-Bond Interaction	π-π Stacking Interaction	Inhibition Rate (%) [2]
A1	Y207-3841	329.78	3.78	−10.40	Tyr368(2) [1], Asn388(2)	Phe426	68.40
A2	ZINC000009481760	395.83	0.38	−9.80	Arg310(2), Ala353, Tyr368(2), Trp385	Phe367	64.25
A3	D306-0032	385.48	3.78	−9.17	Arg310(2), Asn386, Ala353, Tyr368	\	69.15
A4	ZINC000257219502	392.42	0.51	−9.15	Ala353, Tyr368(2)	Phe426	47.16
A5	ZINC000012154664	378.45	0.97	−8.95	Ala353, Tyr368, Asn388	Phe426	20.79
A6	C226-1222	350.33	2.70	−8.85	Ala353, Tyr368, Asn386	Phe426	46.92
A7	ZINC000257316872	430.49	0.86	−8.79	Gly333, Tyr368, Asn388	Phe426	4.07
A8	D665-0380	378.25	1.42	−8.69	Tyr368	Phe426	33.39
A9	ZINC000008892924	478.56	0.84	−8.58	Arg310, Gly333, Asn388	Phe426	11.20

[1] This represents the compound that forms two hydrogen bonds with the same amino acid. [2] The inhibition rate was calculated by taking the average of three parallel experiments at 50 μM.

2.2.2. SBVS Results of nsp16 Inhibitors

In our comprehensive multi-layer virtual screening for nsp16 inhibitors, we focused on two distinct sites, i.e., the SAM binding site and the nsp10–nsp16 interface. For the SAM binding site, we identified eight potential inhibitors B1–B8 (ZINC55183218, ZINC4073149, ZINC95190922, ZINC60349570, ZINC1127559, ZINC65164617, ZINC215527498, ZINC20477654) based on their rank of docking scores and binding interactions (Figures 6, 7, S4 and S7b and Table 2). As shown in Figures 7 and S4, all eight compounds conform to Lipinski's Rule of Five and occupy both the SAM-adenine and SAM-tail cavities. The top three hits B1–B3 (Figure 7, Table 2) with high docking scores less than −8.30 kcal/mol include positively charged amino groups that form salt bridge interactions with Asp6897 and Asp6928 in the SAM-adenine and SAM-tail cavities, respectively, exhibiting a similar binding conformation as SAM and Sinefungin (Figure S2b). In addition, the positively charged amino groups of these three top molecules form hydrogen bond interactions with Gly6871 and Gly6869 in the SAM binding pocket. Especially, B3 significantly exhibited the greatest number of H-bond and salt-bridge interactions with nsp16, indicating that this molecule might be the most potent nsp16 inhibitor. The in vitro 2′-O-MTase inhibition activities of these eight compounds were tested for further validations and comparison.

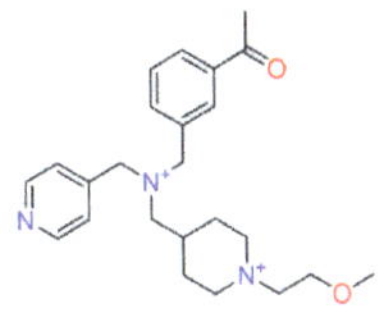

B1

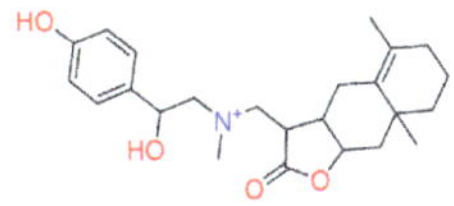

B2

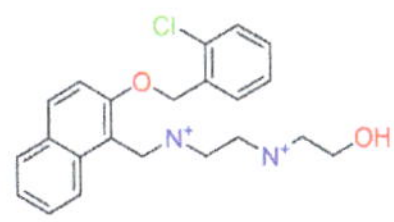

B3

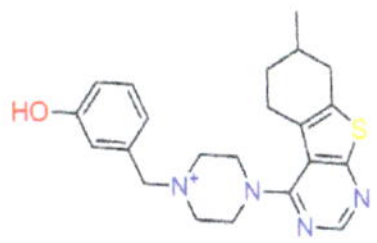

B4

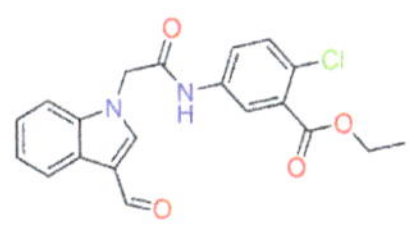

B5

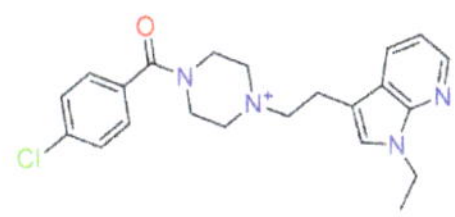

B6

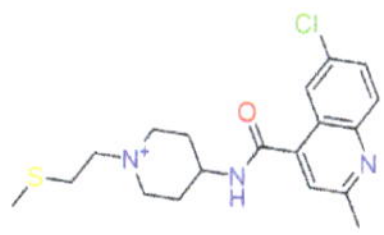

B7

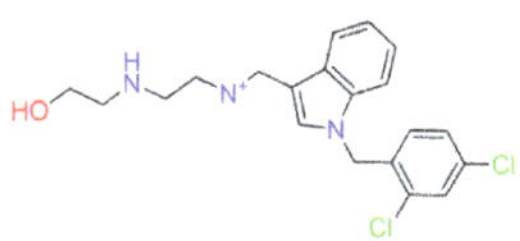

B8

Figure 6. Chemical structures of potential inhibitors B1–B8 targeting nsp16 SAM binding site.

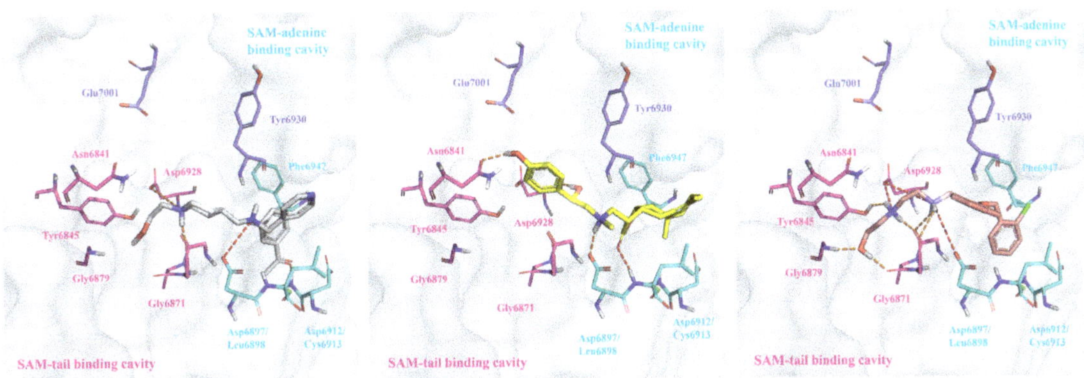

(a) (b) (c)

Figure 7. A three-dimensional view of binding modes of (**a**) B1, (**b**) B2, and (**c**) B3 in the nsp16 SAM binding site. Amino acids in the SAM-adenine and SAM-tail cavities are denoted by cyan and pink sticks, respectively. The hydrogen bonds are denoted by orange dashed lines, and salt bridge interactions are denoted by red dashed lines.

For the screening at the nsp10–nsp16 interface, we identified five compounds C1–C5 (ZINC67911283, ZINC67912643, ZINC95785585, ZINC253387786, and ZINC72320248), which are natural products with large molecular weights and multiple hydroxyl groups. As shown in Figures S5, S6 and S7c and Table S4, they form multiple hydrogen bonds with the predicted key amino acids such as Gln6885, Thr6889, Leu7050, Pro4312, and Glu4313 at the nsp10–nsp16 interface, leading to higher docking scores compared with the selected compounds in the SAM binding site of nsp16. However, they exhibited poor drug-like properties because they contain more than five hydrogen bond donors, contravening Lipinski's Rule of Five for small molecular drugs (Table S4). Thus, we only selected compound C1 with the best binding score and interaction for further in vitro 2′-O-MTase inhibition validation.

Table 2. The SBVS and in vitro validation results of potential inhibitors targeting nsp16 SAM binding site.

Code	Compound	Molecular Weight	LogP	Docking Score (kcal/mol)	H-Bond Interaction	Salt Bridge Interaction	Inhibition Rate (%) [2]
B1	ZINC55183218	397.6	2.50	−8.70	Gly6871, Cys6913	Asp6897, Asp6928,	49.06
B2	ZINC4073149	400.5	3.69	−8.60	Leu6898, Asp6928	Asp6897	48.82
B3	ZINC95190922	386.9	3.52	−8.30	Gly6869(2) [1], Ala6870, Gly6871, Gly6879, Asp6928	Asp6897, Asp6928	54.91
B4	ZINC60349570	395.5	4.63	−8.26	Leu6898, Cys6913, Lys6968, Asp6928	Asp6897	0
B5	ZINC1127559	384.8	3.31	−8.06	Asn6841, Asp6897, Cys6913, Tyr6930	\	26.82
B6	ZINC65164617	397.9	3.42	−7.76	Asp6873, Asp6897	Asp6897	0.78
B7	ZINC215527498	378.9	3.74	−7.57	Tyr6830, Gly6871, Asp6897, Cys6913	Asp6897	0
B8	ZINC20477654	393.3	3.11	−7.51	Gly6869, Ala6870, Gly6879, Asp6928(2)	Asp6928	0

[1] This represents the compound that forms two hydrogen bonds with the same amino acid. [2] The inhibition rate was calculated by taking the average of three parallel experiments at 50 μM.

2.2.3. ADMET Properties of the Selected Potential MTase Inhibitors

The pharmacokinetic properties of the selected potential MTase inhibitors from SBVS were predicted by pkCSM (https://biosig.lab.uq.edu.au/pkcsm/, accessed on 5 March 2023) [39]. The results presented in Tables S5–S7 indicated that the 9 and 8 selected compounds, respectively, targeting the SAM binding site of nsp14 and nsp16 exhibited moderate absorption, distribution, metabolism, and excretion (ADME) properties, so these 17 compounds were all further validated by in vitro MTase inhibition activity testing, whereas the 5 selected natural products targeting the nsp10–nsp16 interface showed poor absorption and metabolism properties, and we only chose compound C1 with the best ADMET properties for further in vitro validation.

2.3. Biochemical Assays for MTase Inhibition Activity

The selected potential inhibitors of nsp14 and nsp16 were validated by in vitro methyltransferase activity testing under a concentration of 50 μM. Consequently, for nsp14, the positive control Sinefungin showed 90.91% of inhibition under 25 μM, and among the 9 selected compounds, 3 (A1–A3) of them exhibited N7-MTase inhibitory rates higher than 60% (Table 1) and were also the top 3 hits in the SBVS. Notably, for nsp16, the positive control Sinefungin showed 86.34% of inhibition under 50 μM, and of the 8 tested compounds binding to the nsp16 SAM site, the top 3 hits (B1–B3) in the virtual screening showed the most potent 2′-O-MTase inhibition activities with inhibitory rates higher than 45% (Table 2), whereas compound ZINC67911283 targeting the nsp10–nsp16 interface only showed a 2′-O-MTase inhibitory rate of 3.72% probably due to its poor drug-like property. As a result, we successfully identified six SARS-CoV-2 MTase inhibitors with moderate activities.

2.4. Molecular Dynamics Simulation

In order to investigate the conformational stability and dynamic features of the six hits, we performed 100 ns MD simulations at 310 K for each system after 700 ps equilibrations. The dynamic parameters such as the root-mean-square deviation (RMSD) of the nsp14 and nsp16 protein backbone and ligands were plotted as a function of time (Figure S8), and the critical distances (Figures S9–S14) that describe the binding interactions between ligands and protein were also analyzed to validate the stability of the binding pose and interactions.

The RMSD trajectories for the protein backbone and ligands shown in Figure S8 indicated that the MD simulations for the six systems reached convergence at around 40 ns, and the dynamic conformations of the proteins and ligands were stable after 40 ns. The critical distance trajectories shown in Figures S9–S14 revealed that the binding poses and key interactions of the six hits in the nsp14/nsp16 protein were almost maintained. For nsp14, all three hits (A1–A3) preserved the conformation of dual substrate occupancy and maintained the hydrogen bond interaction with Tyr368 in the SAM-adenine cavity (Figures S9–S11). For nsp16, the hydrogen bond interactions between the three hits (B1–B3) and the amino acids in the SAM-tail cavity were maintained. Notably, the salt bridge interactions between the positively charged amino groups of the three hits and amino acids Asp6897 or Asp6897 in nsp16 were retained (Figures S12–S14). These results demonstrated the stable conformational stability and dynamic features of the final six hits in the nsp14/nsp16 protein.

3. Discussion

SARS-CoV-2 and its mutants have caused millions of deaths globally. Though effective vaccines and antiviral drugs have been developed, there is evolving resistance to these vaccines and drugs. Consequently, it is desirable to develop antivirals targeting enzymes central to the life cycle of SARS-CoV-2. The methyltransferases nsp14 and nsp16 of SARS-CoV-2 are such enzymes that use SAM as a cofactor to methylate the N7 and 2′-O positions of the 5′-end of viral mRNA to evade the host immune response [40]. Unlike SARS-CoV-2 M^{pro} and RdRp enzymes [41,42] that were developed as investigational drugs through similarities to other viruses, the MTase of nsp14 and nsp16 have seen little research [43]. In addition, most of the reported small-molecule inhibitors of nsp14 or nsp16 are SAM analogs, whose hydrophilicity hinders their ability to cross cell membranes [40]. Thus, it is desirable to discover more nsp14 and nsp16 inhibitors with diverse scaffolds.

In this study, we utilized SBVS to screen out potential small molecular MTase inhibitors of SARS-CoV-2 based on the high-resolution co-crystal structure of nsp14 (PDB ID: 7R2V) and nsp16 (PDB ID: 6WVN). SBVS on the SAM binding sites of the nsp14/nsp16 and nsp10–nsp16 interface was subsequently performed using Glide HTVS, SP, XP docking, and visual inspection of the binding interactions through Schrödinger software. Finally, 9, 8, and 1 compounds targeting SAM binding sites of the nsp14/nsp16 and nsp10–nsp16 interface were screened and validated by in vitro biochemical assays for MTase inhibition activity. The results revealed that 3 potential nsp14 inhibitors A1–A3 exhibited N7-MTase inhibition rates higher than 60%, and 3 compounds B1–B3 targeting the nsp16 SAM binding site showed 2′-O-MTase inhibition rates higher than 45%. Interestingly, the top 3 compounds from SBVS of the nsp14 SAM binding site and the top 3 compounds of the nsp16 SAM binding site were also experimentally validated as the most potent MTase inhibitors, indicating the rationality of our SBVS strategy. Notably, compounds B1–B3 are not SAM analogs but exhibit similar binding modes to SAM, providing new scaffolds for the further study of nsp16 inhibitors. In addition, in the molecular dynamics simulations, we further verified the binding stability of the six identified compounds to their target receptor, illustrating their sustained interaction throughout the simulation trajectories.

There are still limitations in this study. Firstly, the MTase inhibition activities of the top six hits for nsp14 and nsp16 are weak compared with the SAM analog Sinefugin (90.91% inhibition of N7-MTase activity under 25 μM, and 86.34% inhibition of 2′-O-MTase activity under 50 μM). Secondly, the MTase inhibition rate of the tested compound C1 targeting the nsp10–nsp16 interface is only 3.7%. These limitations of SBVS results are probably attributed to the large size of the SAM and interface pockets and the lack of knowledge of crucial residues that determine inhibition potency and selectivity. These factors make the nsp14 and nsp16 targets challenging for docking [43]. In addition, the potency of the nsp10–nsp16 interface as a drug target still requires further experimental validations. Nevertheless, our findings could provide candidates with new scaffolds for the further development of SARS-CoV-2 MTase inhibitors.

4. Materials and Methods

4.1. Processing of nsp14 and nsp16 Structures

The crystal structures of SARS-CoV-2 nsp14 (PDB ID: 7R2V) and SARS-CoV-2 nsp10-nsp16 complexes (PDB ID: 6WVN) for SBVS were downloaded from Protein Data Bank (https://www.rcsb.org/, accessed on 20 October 2022). Prior to SBVS, the structures of SARS-CoV-2 nsp14 and nsp10–nsp16 complexes were prepared using the protein preparation wizard module of Schrödinger software (Release 2019-2, Schrödinger LLC, New York, NY, USA). The protein preparations, including protonation—state adjustment, water removal, disulfide bonds, hydrogen atom and missing heavy atom addition, and structural minimization, were performed by the Maestro module of Schrödinger software.

4.2. Processing of Small Molecules

For SARS-CoV-2 nsp14, we obtained a dataset of 139,000 ligands from the ZINC15 database for virtual screening. These compounds were selected according to the physicochemical properties of the reported active molecules, whose molecular weights and logP values ranged from 375 to 500 and −1 to 1, respectively [17,26,27,29,31,43,44]. Additionally, we selected an extra dataset of 100,000 molecules from ChemDiv (https://www.chemdiv.com/catalog/diversity-libraries, accessed on 30 October 2022/100k Diverse Compounds Pre-Plated Set) to search for inhibitors with diverse scaffolds.

Regarding the SARS-CoV-2 nsp16 SAM binding site, we selected the ligands with a molecular weight of 400 and logP of 4 from the ZINC15 database (https://zinc.docking.org, accessed on 25 October 2022) according to the physicochemical properties of the reported active nsp16 inhibitors [17,18,20,24] and ended up with 210,000 ligands to perform the virtual screening. For the potential binding pocket at the nsp10–nsp16 interface, 237,000 natural products retrieved from the biogenic subset of ZINC15 were collected for SBVS following the virtual screened database used by Mohammad et al. [34].

The downloaded compounds were collected in simplified molecular-input line-entry system (SMILES) format. Then, the LigPrep panel in Maestro was employed for ligand preprocessing which includes (i) an OPLS_2005 force field, (ii) no change for ionization, (iii) a desalt option, (iv) chirality determination from the 3D structure, (v) the generation of one low energy conformer at most per ligand, and (vi) an output in SDF format. The generated 3D conformers of all compounds were subjected to SBVS.

4.3. Structure-Based Virtual Screening

The SBVS parameters in the SAM binding site of SARS-CoV-2 nsp14 and nsp16 were determined by redocking the substrate SAH and SAM into nsp14 and nsp16, respectively, to resume the binding mode and interactions of SAH and SAM in the co-crystal structures of nsp14 (PDB ID: 7R2V) and nsp16 (PDB ID: 6WVN). Consequently, the receptor grids for the SBVS of nsp14 and nsp16 were defined as a 30 Å box centered on the O-atom of residue Asn386 and a 35 Å box centered on the 2′-O atom of the m7GpppA substrate, respectively.

Considering that there is no reported co-crystallized ligand that binds to the SARS-CoV-2 nsp10–nsp16 interface, the SBVS parameters were determined by docking the reported potential natural product inhibitor Genkwanin-6-C-beta-glucopyranoside, which was identified by virtual screening against the nsp10 interface [34], into the predicted binding site of the nsp10-nsp16 interface with different box sizes and docking centers near the key residues using the Glide module of Schrödinger software. As a result, the receptor grid for SBVS was defined as a 30 Å box centered on the O atom of key residue Gln6885 in chain A, because the docking of Genkwanin-6-C-beta-glucopyranoside using this grid generated the best binding affinity and most favorable interactions with nsp10–nsp16 interface residues. All other parameters were kept as the default in Schrödinger.

The virtual screening workflow (HTVS, SP, and XP) of Schrödinger (Maestro 11.6.013) was utilized for the SBVS. Initially, the Glide High-Throughput Virtual Screening (HTVS) mode was employed for the preliminary screening phase, and the top 10% hits with the highest binding scores from HTVS were used for the subsequent filtering by the

Glide Standard Precision (SP) mode. Then, the top 10% hits from SP were utilized for the next round of screening by the Glide Extra Precision (XP) methodology, and the top 20% hits from XP were subjected to the visual inspection screening [36–38]. Finally, the potential small-molecule inhibitors exhibited similar or more favorable binding interactions compared with Sinefungin, which were selected and purchased from TargetMol (https://www.tsbiochem.com/, accessed on 20 February 2023).

4.4. Biochemical Assays

4.4.1. RNA Substrate Preparation

The RNA substrate of 5′-terminal 259 nucleotides (ATP as the viral initial nucleotide) of the SARS-CoV-2 genome (uncapped SARS-CoV-2 RNA) was in vitro-transcribed from PCR products by using the MEGAscript Kit (Ambion, Austin, TX, USA) as described in our previous work [45]. By using the vaccinia virus capping enzyme system (Novoprotein, Suzhou, China), the transcribed RNAs were capped/methylated to form the GpppG/A and m7GpppG/A-capped RNAs in the presence or absence of the methyl donor SAM. Primers used for the synthesis of RNA substrates were as follows—Forward-5′: TAATACGACTCACTATTAGATTAAAGGTTTATACCTTCCCAGG, Reverse-5′: CTTTCG-GTCACACCCGGAC.

4.4.2. Protein Expression and Purification

The coding sequences of SARS-CoV nsp10, 16; SARS-CoV-2 nsp10, 14, 16; and mutants were cloned into a pET32a vector with the His tag. *E. coli* BL21 (DE3) cells were transformed with the respective plasmid and the recombinant protein was induced with 0.4 mM isopropyl β-d-thiogalactopyranoside (IPTG) at 16 °C for 12–16 h. The cells were harvested by centrifugation, and the pellets were resuspended in lysis buffer (50 mM Tris–HCl, pH 8.0, 300 mM NaCl, 10% glycerol, and 5 mM $MgCl_2$). The cells were then disrupted by a high-pressure cracker (UH-24, Union-biotech, Shanghai, China), and cell debris was removed by centrifugation. pET32a-His6-nsp10, 14, and 16 were purified with nickel-nitrilotriacetic acid (Ni-NTA, Shanghai, China) resin (GenScript, Piscataway, NJ, USA) as described previously [45,46].

4.4.3. Radioactive Biochemical Assays for MTase Activity

SARS-CoV-2 nsp14 and nsp16/10 inhibition assays of the final selected compound were carried out in a 30 μL reaction mixture [40 mM Tris-HCl (pH 7.5), 2 mM $MgCl_2$, 2 mM DTT, 40 units RNase inhibitor, 0.01 mM SAM], with 0.5 μCi of S-adenosyl [methyl-3H] methionine (67.3 Ci/mmol, 0.5 μCi/mL), 1 μg of purified proteins, and 2 μg of m7GpppA RNA substrates at 37 °C for 1.5 h. The 3H-labeled product was isolated in small DEAE-Sephadex columns and quantitated by liquid scintillation.

4.5. Molecular Dynamics Simulations

The molecular dynamics (MD) simulation was performed by the Amber 18 software package installed on a Linux platform. The high-performance server cluster on the platform was composed of two Intel® Xeon® Platinum 8176 CPU (Intel, Santa Clara, CA, USA) processors accelerated by two NVIDIA Tesla V100 SXM2 GPUs (NVIDIA, Santa Clara, CA, USA). The FF14SB force field and GAFF2 force field were applied to proteins and ligands, respectively. Throughout the simulation, the coordination distances of the zinc ions including ZN-S (2.40 Å) and ZN-N (2.10 Å) were restrained by a binding constant of 25 kcal mol^{-1} $Å^{-2}$ [47]. Next, the charges of the 6 compounds were calculated and assigned by Gaussian16 and the Restrained Electro Static Potential (RESP) module in Amber18 software packages. The complexes were then solvated into the pre-equilibrated TIP3P water under periodic boundary conditions using a cubic box model with a 15 Å buffer distance, and 15 and 7 Cl^- ions were added to neutralize the system of nsp14 and nsp16, respectively. Subsequent energy minimizations and MD equilibrium simulations followed a similar protocol to our previous studies [48,49]. We first performed 2500 steps of steepest descent minimization followed by a 2500-cycle conjugate gradient minimization by restraining the protein and ligand with a force constant of 50 kcal·mol^{-1}·Å^{-2}. Then, we performed a 100 ps NVT equilibration simulation (T = 10 K) followed by another NPT (P = 1 atm) equilibration simulation with the restraint force constant gradually decreased to 25 kcal·mol^{-1}·Å^{-2}. Next, we performed a 200 ps temperature annealing NVT simulation (T was raised from 10 K to 310 K) and a 100 ps NPT simulation with the restraint force constant reduced to 10 kcal·mol^{-1}·Å^{-2}. Then, we performed two sequential 100 ps NPT simulations with reduced restraint force constants of 1 and 0.1 kcal·mol^{-1}·Å^{-2}, respectively. Finally, a 100 ns production NPT (T = 310 K and P = 1 atm) simulation was carried out without any restraints. In all MD simulations, the SHAKE algorithm was utilized to constrain the bond length [50], and a 10 Å cutoff was used for both short-range and van der Waals (vdW) interactions. The integration was kept with a 2-fs step. The minimizations and equilibrations were carried out by the sander module, and the three independent 100 ns production MD simulations were performed by the PMEMD.CUDA module in Amber18 [51]. The trajectories were analyzed using the CPPTRAJ package in Amber18 [52].

5. Conclusions

In summary, SBVS was performed to explore potential SARS-CoV-2 nsp14 and nsp16 methyltransferase inhibitors. The virtual screening workflow (HTVS, SP, XP) in Schrödinger software combined with visual inspection of the binding interactions were used for our screening of 349,000 compounds from the ZINC database and 100,000 compounds from the ChemDiv database. Consequently, the top 9 and 8 hits targeting SAM binding sites of nsp14 and nsp16, respectively, with the best binding affinities and most favorable interactions were filtered out for further in vitro MTase inhibition activity validation. Finally, three potential inhibitors A1–A3 of nsp14 were identified which exhibited over 60% of inhibition of N7-MTase activity under a concentration of 50 μM. Moreover, three molecules B1–B3 surpassing 45% of inhibition of 2′-O-MTase activity at the concentration of 50 μM were identified as potential inhibitors for nsp16. These findings could provide potential lead compounds for the rational drug design of SARS-CoV-2.

Supplementary Materials: The following supporting information can be downloaded at https://www.mdpi.com/article/10.3390/molecules29102312/s1. Table S1: The X-ray data analysis of the SRAS-CoV-2 nsp14 structures reported in the PDB database. Table S2: The X-ray data analysis of the SRAS-CoV-2 nsp10–nsp16 complexes reported in the PDB database. Figure S1: (a) The DoGSiteScorer-predicted binding pocket (shown by surface) at the nsp10–nsp16 interface (6WVN) with a druggability score of 0.81 and volume of 589.38 Å^3; (b) schematic of the non-bonded interactions between interface residues of nsp16 (Chain A) and nsp10 (Chain B) extracted by PDBsum prot-prot analysis. Table S3. The amino acid residues positioned in the DoGSiteScorer predicted binding pocket of nsp10–nsp16 interface (key residues are labeled by bold). Figure S2: (a) A three-dimensional view of the binding interactions of Sinefungin in the nsp14 SAM binding site; (b) a three-dimensional view of the binding interactions of Sinefungin in the nsp16 SAM binding site. Figure S3: A three-dimensional view of the binding interaction of A4–A9 (a–f) in the nsp14 SAM binding site. Figure S4: A three-dimensional view of the binding interaction of B4–B8 (a–e) in the nsp16 SAM binding site. Figure S5: Chemical structures of nsp10–nsp16 interface site inhibitors. Figure S6: A three-dimensional view of the binding interaction of C1–C5 (a–e) at the nsp10–nsp16 interface. Table S4: Molecular docking results of nsp10–nsp16 interface site inhibitors. Figure S7: Docking score distribution of glide HTVS, SP, and XP docking steps in VS of compounds targeting the SAM binding sites of nsp14 (a) and nsp16 (b) and the nsp10–nsp16 interface (c). Table S5: Predicted ADMET properties of SARS-CoV-2 nsp14 inhibitors by pkCSM. Table S6: Predicted ADMET properties of SARS-CoV-2 nsp16 inhibitors by pkCSM. Table S7: Predicted ADMET properties of SARS-CoV-2 nsp10–nsp16 interface inhibitors by pkCSM. Figure S8: The RMSD of protein backbone (a) and ligands (b) as a function of time for the 6 hits bound to SARS-CoV-2 nsp14 and nsp16 SAM binding sites in the 100 ns MD simulations. Figures S9–S14: The predicted 3D binding mode of compounds (A1–A3, B1–B3) bound to SARS-CoV-2 nsp14/nsp16, and the distances that describe the binding interactions between the ligand and protein as a function of time in the 100 ns MD simulation.

Author Contributions: Data curation, K.W. and Y.G.; formal analysis, K.W. and Y.G.; funding acquisition, J.L.; methodology, K.W. and Y.G.; project administration, J.L.; validation, T.X., W.H., D.G. and L.C.; writing—original draft preparation, K.W. and Y.G.; writing—review and editing, W.H. and J.L. All authors have read and agreed to the published version of the manuscript.

Funding: This work has been supported by the National Key Research and Development Program of China (2023YFF1204900, 2023YFF1204902), and the Guangdong-Hong Kong Technology Cooperation Funding Scheme (2023A0505010015).

Institutional Review Board Statement: Not applicable.

Informed Consent Statement: Not applicable.

Data Availability Statement: All data are included in the article and Supplementary Materials.

Conflicts of Interest: Author Deyin Guo was employed by the company Bio-Island. The remaining authors declare that the research was conducted in the absence of any commercial or financial relationships that could be construed as a potential conflict of interest.

References

1. Gogoi, B.; Chowdhury, P.; Goswami, N.; Gogoi, N.; Naiya, T.; Chetia, P.; Mahanta, S.; Chetia, D.; Tanti, B.; Borah, P.; et al. Identification of potential plant-based inhibitor against viral proteases of SARS-CoV-2 through molecular docking, mm-pbsa binding energy calculations and molecular dynamics simulation. *Mol. Divers.* **2021**, *25*, 1963–1977. [CrossRef] [PubMed]
2. Maunder, R.; Hunter, J.; Vincent, L.; Bennett, J.; Peladeau, N.; Leszcz, M.; Sadavoy, J.; Verhaeghe, L.M.; Steinberg, R.; Mazzulli, T. The immediate psychological and occupational impact of the 2003 SARS outbreak in a teaching hospital. *Cmaj* **2003**, *168*, 1245–1251. [PubMed]
3. Rabaan, A.A.; Al-Ahmed, S.H.; Haque, S.; Sah, R.; Tiwari, R.; Malik, Y.S.; Dhama, K.; Yatoo, M.I.; Bonilla-Aldana, D.K.; Rodriguez-Morales, A.J. SARS-CoV-2, SARS-CoV, and MERS-CoV: A comparative overview. *Infez. Med.* **2020**, *28*, 174. [PubMed]
4. Zhou, M.; Zhang, X.; Qu, J. Coronavirus disease 2019 (COVID-19): A clinical update. *Front. Med.* **2020**, *14*, 126–135. [CrossRef] [PubMed]
5. Carmen Espinosa-Gongora, C.B.M.R. Early detection of the emerging SARS-CoV-2 ba.2.86 lineage through integrated genomic surveillance of wastewater and COVID-19 cases in sweden, weeks 31 to 38 2023. *Eurosurveillance* **2023**, *46*, 2300595. [CrossRef] [PubMed]
6. Li, G.; De Clercq, E. Therapeutic options for the 2019 novel coronavirus (2019-ncov). *Nat. Rev. Drug Discov.* **2020**, *19*, 149–150. [CrossRef]
7. Yu, B.; Chang, J. Azvudine (fnc): A promising clinical candidate for COVID-19 treatment. *Signal Transduct. Target. Ther.* **2020**, *5*, 236. [CrossRef] [PubMed]
8. Zhang, W.; Li, L.; Zhou, Z.; Liu, Q.; Wang, G.; Liu, D. Cost-effectiveness of paxlovid in reducing severe COVID-19 and mortality in china. *Front. Public Health* **2023**, *11*, 1174879. [CrossRef]
9. Tran, N.; Khoi Quan, N.; Tran, V.P.; Nguyen, D.; Tao, N.P.H.; Linh, N.N.H.; Tien Huy, N. Molnupiravir as the COVID-19 panacea: False beliefs in low- and middle-income countries. *Pathog. Glob. Health* **2023**, *117*, 525–526. [CrossRef]
10. Gurung, A.B. In silico structure modelling of SARS-CoV-2 nsp13 helicase and nsp14 and repurposing of fda approved antiviral drugs as dual inhibitors. *Gene Rep.* **2020**, *21*, 100860. [CrossRef]
11. Finkel, Y.; Mizrahi, O.; Nachshon, A.; Weingarten-Gabbay, S.; Morgenstern, D.; Yahalom-Ronen, Y.; Tamir, H.; Achdout, H.; Stein, D.; Israeli, O.; et al. The coding capacity of SARS-CoV-2. *Nature* **2021**, *589*, 125–130. [CrossRef] [PubMed]
12. Ramanathan, A.; Robb, G.B.; Chan, S. Mrna capping: Biological functions and applications. *Nucleic Acids Res.* **2016**, *44*, 7511–7526. [CrossRef] [PubMed]
13. Krafcikova, P.; Silhan, J.; Nencka, R.; Boura, E. Structural analysis of the SARS-CoV-2 methyltransferase complex involved in rna cap creation bound to sinefungin. *Nat. Commun.* **2020**, *11*, 3717. [CrossRef] [PubMed]
14. Lin, S.; Chen, H.; Ye, F.; Chen, Z.; Yang, F.; Zheng, Y.; Cao, Y.; Qiao, J.; Yang, S.; Lu, G. Crystal structure of SARS-CoV-2 nsp10/nsp16 2′-o-methylase and its implication on antiviral drug design. *Signal Transduct. Target. Ther.* **2020**, *5*, 131. [CrossRef] [PubMed]
15. Rogstam, A.; Nyblom, M.; Christensen, S.; Sele, C.; Talibov, V.O.; Lindvall, T.; Rasmussen, A.A.; André, I.; Fisher, Z.; Knecht, W.; et al. Crystal structure of non-structural protein 10 from severe acute respiratory syndrome coronavirus-2. *Int. J. Mol. Sci.* **2020**, *21*, 7375. [CrossRef] [PubMed]
16. Bouvet, M.; Lugari, A.; Posthuma, C.C.; Zevenhoven, J.C.; Bernard, S.; Betzi, S.; Imbert, I.; Canard, B.; Guillemot, J.; Lécine, P.; et al. Coronavirus nsp10, a critical co-factor for activation of multiple replicative enzymes. *J. Biol. Chem.* **2014**, *289*, 25783–25796. [CrossRef]
17. Bobrovs, R.; Kanepe, I.; Narvaiss, N.; Patetko, L.; Kalnins, G.; Sisovs, M.; Bula, A.L.; Grinberga, S.; Boroduskis, M.; Ramata-Stunda, A.; et al. Discovery of SARS-CoV-2 nsp14 and nsp16 methyltransferase inhibitors by high-throughput virtual screening. *Pharmaceuticals* **2021**, *14*, 1243. [CrossRef]

18. Klima, M.; Khalili Yazdi, A.; Li, F.; Chau, I.; Hajian, T.; Bolotokova, A.; Kaniskan, H.Ü.; Han, Y.; Wang, K.; Li, D.; et al. Crystal structure of SARS-CoV-2 nsp10–nsp16 in complex with small molecule inhibitors, ss148 andwz16. *Protein Sci.* **2022**, *31*, e4395. [CrossRef]
19. Kottur, J.; Rechkoblit, O.; Quintana-Feliciano, R.; Sciaky, D.; Aggarwal, A.K. High-resolution structures of the SARS-CoV-2 n7-methyltransferase inform therapeutic development. *Nat. Struct. Mol. Biol.* **2022**, *29*, 850–853. [CrossRef]
20. Sulimov, A.; Kutov, D.; Ilin, I.; Xiao, Y.; Jiang, S.; Sulimov, V. Novel inhibitors of 2′-o-methyltransferase of the SARS-CoV-2 coronavirus. *Molecules* **2022**, *27*, 2721. [CrossRef]
21. Shang, Z.; Chan, S.Y.; Liu, W.J.; Li, P.; Huang, W. Recent insights into emerging coronavirus: SARS-CoV-2. *Acs Infect. Dis.* **2021**, *7*, 1369–1388. [CrossRef]
22. Schöning-Stierand, K.; Diedrich, K.; Fährrolfes, R.; Flachsenberg, F.; Meyder, A.; Nittinger, E.; Steinegger, R.; Rarey, M. Proteinsplus: Interactive analysis of protein–ligand binding interfaces. *Nucleic Acids Res.* **2020**, *48*, W48–W53. [CrossRef]
23. Williams, C.J.; Headd, J.J.; Moriarty, N.W.; Prisant, M.G.; Videau, L.L.; Deis, L.N.; Verma, V.; Keedy, D.A.; Hintze, B.J.; Chen, V.B.; et al. Molprobity: More and better reference data for improved all-atom structure validation. *Protein Sci.* **2018**, *27*, 293–315. [CrossRef] [PubMed]
24. Rosas-Lemus, M.; Minasov, G.; Shuvalova, L.; Inniss, N.L.; Kiryukhina, O.; Brunzelle, J.; Satchell, K. High-resolution structures of the SARS-CoV-2 2′-o-methyltransferase reveal strategies for structure-based inhibitor design. *Sci. Signal.* **2020**, *13*, eabe1202. [CrossRef] [PubMed]
25. Czarna, A.; Plewka, J.; Kresik, L.; Matsuda, A.; Karim, A.; Robinson, C.; O Byrne, S.; Cunningham, F.; Georgiou, I.; Wilk, P.; et al. Refolding of lid subdomain of SARS-CoV-2 nsp14 upon nsp10 interaction releases exonuclease activity. *Structure* **2022**, *30*, 1050–1054. [CrossRef] [PubMed]
26. Ahmed-Belkacem, R.; Hausdorff, M.; Delpal, A.; Sutto-Ortiz, P.; Colmant, A.M.G.; Touret, F.; Ogando, N.S.; Snijder, E.J.; Canard, B.; Coutard, B.; et al. Potent inhibition of SARS-CoV-2 nsp14n 7-methyltransferase by sulfonamide-based bisubstrate analogues. *J. Med. Chem.* **2022**, *65*, 6231–6249. [CrossRef] [PubMed]
27. Jung, E.; Soto-Acosta, R.; Xie, J.; Wilson, D.J.; Dreis, C.D.; Majima, R.; Edwards, T.C.; Geraghty, R.J.; Chen, L. Bisubstrate inhibitors of severe acute respiratory syndrome coronavirus-2 nsp14 methyltransferase. *Acs Med. Chem. Lett.* **2022**, *13*, 1477–1484. [CrossRef] [PubMed]
28. Amador, R.; Delpal, A.; Canard, B.; Vasseur, J.J.; Decroly, E.; Debart, F.; Clave, G.; Smietana, M. Facile access to 4′-(n-acylsulfonamide) modified nucleosides and evaluation of their inhibitory activity against SARS-CoV-2 rna cap n7-guanine-methyltransferase nsp14. *Org. Biomol. Chem.* **2022**, *20*, 7582–7586. [CrossRef] [PubMed]
29. Bobileva, O.; Bobrovs, R.; Sirma, E.E.; Kanepe, I.; Bula, A.L.; Patetko, L.; Ramata-Stunda, A.; Grinberga, S.; Jirgensons, A.; Jaudzems, K. 3-(adenosylthio)benzoic acid derivatives as SARS-CoV-2 nsp14 methyltransferase inhibitors. *Molecules* **2023**, *28*, 768. [CrossRef]
30. Samrat, S.K.; Bashir, Q.; Zhang, R.; Huang, Y.; Liu, Y.; Wu, X.; Brown, T.; Wang, W.; Zheng, Y.G.; Zhang, Q.; et al. A universal fluorescence polarization high throughput screening assay to target the sam-binding sites of SARS-CoV-2 and other viral methyltransferases. *Emerg. Microbes Infect.* **2023**, *12*, 2204164. [CrossRef]
31. Bobiļeva, O.; Bobrovs, R.; Kaņepe, I.; Patetko, L.; Kalniņš, G.; Aišovs, M.; Bula, A.L.; Grīnberga, S.; Borodušķis, M.; Ramata-Stunda, A.; et al. Potent SARS-CoV-2 mrna cap methyltransferase inhibitors by bioisosteric replacement of methionine in sam cosubstrate. *Acs Med. Chem. Lett.* **2021**, *12*, 1102–1107. [CrossRef]
32. Sk, M.F.; Jonniya, N.A.; Roy, R.; Poddar, S.; Kar, P. Computational investigation of structural dynamics of SARS-CoV-2 methyltransferase-stimulatory factor heterodimer nsp16/nsp10 bound to the cofactor sam. *Front. Mol. Biosci.* **2020**, *7*, 590165. [CrossRef]
33. Aldahham, B.J.M.; Al-Khafaji, K.; Saleh, M.Y.; Abdelhakem, A.M.; Alanazi, A.M.; Islam, M.A. Identification of naphthyridine and quinoline derivatives as potential nsp16-nsp10 inhibitors: A pharmacoinformatics study. *J. Biomol. Struct. Dyn.* **2022**, *40*, 3899–3906. [CrossRef] [PubMed]
34. Mohammad, A.; Alshawaf, E.; Marafie, S.K.; Abu-Farha, M.; Al-Mulla, F.; Abubaker, J. Molecular simulation-based investigation of highly potent natural products to abrogate formation of the nsp10–nsp16 complex of SARS-CoV-2. *Biomolecules* **2021**, *11*, 573. [CrossRef] [PubMed]
35. Volkamer, A.; Kuhn, D.; Grombacher, T.; Rippmann, F.; Rarey, M. Combining global and local measures for structure-based druggability predictions. *J. Chem. Inf. Model.* **2012**, *52*, 360–372. [CrossRef] [PubMed]
36. Gupta, Y.; Dawid, M.; Samantha, E.Z.; Krysten, A.J. Bisindolylmaleimide ix: A novel anti-SARS-CoV-2 agent targeting viral main protease 3clpro demonstrated by virtual screening pipeline and in-vitro validation assays. *Methods* **2021**, *195*, 57–71. [CrossRef] [PubMed]
37. Ruiz, V.; Czyzyk, D.J.; Valhondo, M.; Jorgensen, W.L.; Anderson, K.S. Novel allosteric covalent inhibitors of bifunctional cryptosporidium hominis ts-dhfr from parasitic protozoa identified by virtual screening. *Bioorg. Med. Chem. Lett.* **2019**, *29*, 1413–1418. [CrossRef] [PubMed]
38. Zong, K.; Xu, L.; Hou, Y.; Zhang, Q.; Che, J.; Zhao, L.; Li, X. Virtual screening and molecular dynamics simulation study of influenza polymerase pb2 inhibitors. *Molecules* **2021**, *26*, 6944. [CrossRef] [PubMed]
39. Pires, D.E.V.; Blundell, T.L.; Ascher, D.B. Pkcsm: Predicting small-molecule pharmacokinetic and toxicity properties using graph-based signatures. *J. Med. Chem.* **2015**, *58*, 4066–4072. [CrossRef]

40. Kottur, J.; White, K.M.; Rodriguez, M.L.; Rechkoblit, O.; Quintana-Feliciano, R.; Nayar, A.; García-Sastre, A.; Aggarwal, A.K. Structures of SARS-CoV-2 n7-methyltransferase with dot1l and prmt7 inhibitors provide a platform for new antivirals. *PLoS Pathog.* **2023**, *19*, e1011546. [CrossRef]
41. Beigel, J.H.; Tomashek, K.M.; Dodd, L.E.; Mehta, A.K.; Zingman, B.S.; Kalil, A.C.; Hohmann, E.; Chu, H.Y.; Luetkemeyer, A.; Kline, S.; et al. Remdesivir for the treatment of COVID-19—final report. *N. Engl. J. Med.* **2020**, *383*, 1813–1826. [CrossRef]
42. Tarannum, H.; Rashmi, K.M.; Nandi, S. Exploring the SARS-CoV-2 main protease (m(pro)) and rdrp targets by updating current structure-based drug design utilizing co-crystals to combat COVID-19. *Curr. Drug Targets* **2022**, *23*, 802–817. [CrossRef]
43. Singh, I.; Li, F.; Fink, E.A.; Chau, I.; Li, A.; Rodriguez-Hernández, A.; Glenn, I.; Zapatero-Belinchón, F.J.; Rodriguez, M.L.; Devkota, K.; et al. Structure-based discovery of inhibitors of the SARS-CoV-2 nsp14 n7-methyltransferase. *J. Med. Chem.* **2023**, *66*, 7785–7803. [CrossRef]
44. Basu, S.; Mak, T.; Ulferts, R.; Wu, M.; Deegan, T.; Fujisawa, R.; Tan, K.W.; Lim, C.T.; Basier, C.; Canal, B.; et al. Identifying SARS-CoV-2 antiviral compounds by screening for small molecule inhibitors of nsp14 rna cap methyltransferase. *Biochem. J.* **2021**, *478*, 2481–2497. [CrossRef]
45. Chen, Y.; Cai, H.; Pan, J.; Xiang, N.; Tien, P.; Ahola, T.; Guo, D. Functional screen reveals SARS coronavirus nonstructural protein nsp14 as a novel cap n7 methyltransferase. *Proc. Natl. Acad. Sci. USA* **2009**, *106*, 3484–3489. [CrossRef]
46. Chen, Y.; Su, C.; Ke, M.; Jin, X.; Xu, L.; Zhang, Z.; Wu, A.; Sun, Y.; Yang, Z.; Tien, P.; et al. Biochemical and structural insights into the mechanisms of SARS coronavirus rna ribose 2′-o-methylation by nsp16/nsp10 protein complex. *PLoS Pathog.* **2011**, *7*, e1002294. [CrossRef]
47. Xu, M.; Zhu, T.; Zhang, J.Z.H. Automatically constructed neural network potentials for molecular dynamics simulation of zinc proteins. *Front. Chem.* **2021**, *9*, 692200. [CrossRef]
48. Lei, J.; Zhou, Y.; Xie, D.; Zhang, Y. Mechanistic insights into a classic wonder drug—Aspirin. *J. Am. Chem. Soc.* **2015**, *137*, 70–73. [CrossRef]
49. Lei, J.; Sheng, G.; Cheung, P.P.; Wang, S.; Li, Y.; Gao, X.; Zhang, Y.; Wang, Y.; Huang, X. Two symmetric arginine residues play distinct roles inthermus thermophilus argonaute dna guide strand-mediated dna target cleavage. *Proc. Natl. Acad. Sci. USA* **2019**, *116*, 845–853. [CrossRef]
50. Kräutler, V.; van Gunsteren, W.F.; Hünenberger, P.H. A fast shake algorithm to solve distance constraint equations for small molecules in molecular dynamics simulations. *J. Comput. Chem.* **2001**, *22*, 501–508. [CrossRef]
51. Peramo, A. Solvated and generalised born calculations differences using gpu cuda and multi-cpu simulations of an antifreeze protein with amber. *Mol. Simulat.* **2016**, *45*, 1263–1273. [CrossRef]
52. Roe, D.R.; Cheatham, T.E. Ptraj and cpptraj: Software for processing and analysis of molecular dynamics trajectory data. *J. Chem. Theory Comput.* **2013**, *9*, 3084–3095. [CrossRef]

molecules

MDPI

Article

NIMO: A Natural Product-Inspired Molecular Generative Model Based on Conditional Transformer

Xiaojuan Shen [1], Tao Zeng [1], Nianhang Chen [1], Jiabo Li [2,*] and Ruibo Wu [1,*]

1 School of Pharmaceutical Sciences, Sun Yat-sen University, Guangzhou 510006, China; shenxj9@mail2.sysu.edu.cn (X.S.); zengt28@mail2.sysu.edu.cn (T.Z.); nhchen95@gmail.com (N.C.)
2 ChemXAI Inc., 53 Barry Lane, Syosset, NY 11791, USA
* Correspondence: jiabo.li@chemxai.com (J.L.); wurb3@mail.sysu.edu.cn (R.W.)

Abstract: Natural products (NPs) have diverse biological activity and significant medicinal value. The structural diversity of NPs is the mainstay of drug discovery. Expanding the chemical space of NPs is an urgent need. Inspired by the concept of fragment-assembled pseudo-natural products, we developed a computational tool called NIMO, which is based on the transformer neural network model. NIMO employs two tailor-made motif extraction methods to map a molecular graph into a semantic motif sequence. All these generated motif sequences are used to train our molecular generative models. Various NIMO models were trained under different task scenarios by recognizing syntactic patterns and structure–property relationships. We further explored the performance of NIMO in structure-guided, activity-oriented, and pocket-based molecule generation tasks. Our results show that NIMO had excellent performance for molecule generation from scratch and structure optimization from a scaffold.

Keywords: natural products; molecular generation; deep learning; fragmentation; transformer

1. Introduction

Natural products (NPs) are derived from evolutionary selection over millions of years to bind to biological macromolecules and therefore possess important biological activity and pharmaceutical value [1]. With the rapid development of pharmacology and synthesis, more and more natural products are coming to our attention as an important source of new bioactive compounds with novel molecular scaffolds [2]. According to a comprehensive study, 6% of all small-molecule drugs approved between 1981 and 2014 are unaltered NPs, 26% are NP derivatives, and 32% are NP mimetics and/or contain an NP pharmacophore [3]. Judging by the average number of natural product-derived fragments (NPFs) in approved drugs since 1939, pharmaceutical drug discovery programs continue to benefit from the use of NPFs [4,5]. As far as we know, NPs have high diversity and structural complexity, such as a high fraction of sp^3 carbon atoms, stereogenic centers, and diverse ring systems, which make them a largely unexplored chemical space and able to be widely incorporated into the pipelines of drug design on a large scale [6].

Inspired by pre-validated NP repositories in nature, e.g., biology-oriented synthesis [7,8] (BIOS) and pseudo-natural product (pseudo-NP) strategy [9,10] (Figure 1), many novel biologically relevant compounds are designed and synthesized. For BIOS, a conserved core scaffold is identified during the lead identification phase and often kept throughout the rest of the compound collection design. Scaffold synthesis and decoration following BIOS could yield new compounds. On the other hand, for pseudo-NPs, the biological relevance of NPs merges with the rapid accessibility through fragmentation and reassembly, going beyond existing NP scaffolds into unexplored chemical space to overcome the limitations of BIOS [11,12]. Overall, dynamic combinatorial chemistry [13] plays an important role in natural product research.

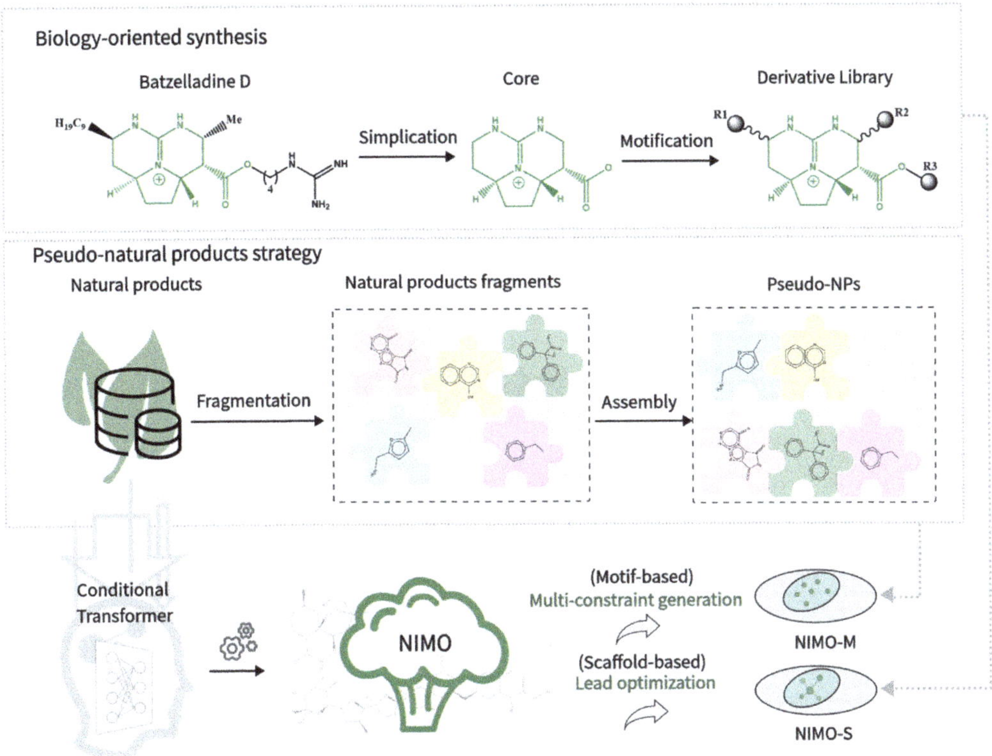

Figure 1. The two bio-inspired ideas integrated into our developed NIMO.

Computationally, many de novo molecular generative models are aimed at generating compound structures with desired physicochemical and bioactivity properties or even multi-objective optimization [14–16]. In terms of granularity, SMILES (simplified molecular-input line-entry system) strings are often adopted as a molecular representation due to their simplicity. For instance, MCMG [17] is a multi-constraint molecular generation approach based on a transformer [18] for de novo drug design. QCMG [19] is a quasi-biogenic molecule generator with recurrent neural networks. However, SMILES-based models often undergo substantial changes during sequential extensions [20]. For example, two molecules with similar chemical structures may be encoded into significantly different SMILES strings. The impact of this characteristic on the structures of polycyclic complex natural products has not been adequately assessed. Some latecomers of graph generation schemes, such as JT-VAE [21], which generates graphs in a motif-by-motif manner rather than node-by-node, are employed to obviate chemically invalid intermediates. Such generators belong to the same fragment-based model as FBMG [22] with respect to the granularity of their applied molecular representation. In this context, there are still relatively few generation models focused on natural products. Given that new chemical entities are typically derived from structural modification of active natural products obtained through screening techniques [23–25], there is a substantial demand for multi-objective structural optimization in natural product-derived models, such as the derivation of the scaffold, in addition to de novo generation [26]. In particular, fragment-based paradigms are thought to be suitable for this real scenario [27–29]. The scaffold-based models typically support scaffold as the initial seed of the generative procedure [30,31]. Last but not least, natural products often feature biologically relevant molecular scaffolds and pharmacophore patterns [4,32]. However, the current molecular generators ignore this critical transfer of relevant structural features to NP-inspired compound libraries.

To our knowledge, previous generative models appear to be confined in the face of the following critical challenges towards natural products: (1) manipulability for complex natural product structures including stereo information, (2) multi-objective structural optimization, and (3) inheritance of biological relevance from natural products. Thus, flexible generative models are needed to complement routine design strategies in real scenarios. In this work, we develop a natural product-inspired molecular generative model (called NIMO for short) based on transformer architecture, in which the rich semantic information among motifs from the aforementioned BIOS and pseudo-NP strategies is learned, and construct the NIMO-S and NIMO-M models, respectively, for generating natural product-like molecular structures that comply with the expected criteria. Specifically, NIMO-M is a generic model for molecular generation with multi-constraint and novel motifs, while NIMO-S is a scaffold-based model for lead optimization that specifies a central scaffold. Furthermore, we thoroughly investigated NIMO in the multi-objective molecular generation tasks and show that NIMO excels in several classical methods in three practical tasks, covering structure-guided, activity-oriented, and pocket-based molecule generation.

2. Results and Discussion

2.1. Model Evaluation

We first trained and sampled 5000 molecules under evaluation setting 1 (see Section 4.3). As shown in Figure 2, the reconstructed chemical spaces of NIMO-S and NIMO-M exhibited a spatial contraction toward the desired properties (QED, logP, and SAS) in contrast to the native chemical space of the NPs. It should be noted that the Mw (molecular weight) distribution was optimized due to the correlation with QED. The statistics showed similar and slightly concentrated molecular property distributions for HBD, HBA, and RB. To assess whether the method can capture the intrinsic structure features of natural products, NP-likeness [33] was introduced as a measure of similarity to the NP molecules, and it showed that both models could generate molecules with more preferred features of NP-like compounds than the synthetic molecules in ZINC [34].

We reported the benchmark studies of 5000 generated molecules under evaluation setting 2 (see Section 4.4). Here, we used SMILES-based models (MCMG and QBMG) and a fragment-based model (FBMG) as the baselines, and the conditional metrics and MOSES [35] metrics were utilized as comprehensive evaluation benchmarks. Table 1 illustrates that all models except FBMG performed pretty well for the validity rate (above 90%). In particular, NIMO not only had significantly higher fragmentation efficiency than FBMG but also yielded a smaller motif size and lower motif weight (Figure S1, Table S1). Because FBMG had difficulties in proposing valid molecules, in addition to its inability to handle stereochemical information and multi-constraint generation, it was not considered in later analysis. On the other side, the validity of the molecules generated by NIMO-M dropped down to 75.12% when the motif information was removed from the training set. This verifies that additional motif information was conducive to model training, thus guaranteeing model reliability and improving training efficiency. Details of the ablation experiment are available in Figure S2. All models scored high on the uniqueness indicator.

In terms of novelty, NIMO-S offered the best performance, while NIMO-M underwent the sharpest drop, to 61.0%. The relatively underperforming MCMG achieved 65.7%. According to MOSES metrics, NIMO-M and MCMG consistently performed well in terms of FCD metrics, which were related to chemical and biological properties. Compared to the SNN metrics, the structures generated by NIMO-S were the furthest from the manifold of the training set. The Frag and Scaf metrics compared molecular similarities at the substructure level. Note that the metric calculation method applied the Bemis–Murcko scaffolds, which partially overlapped with our motif extraction method, so it was comprehensible for the enhancement of Frag metrics of NIMO-S. All models showed roughly the same IntDiv of the generated molecules, indicating the diversity of the generated molecules.

NIMO-M achieved the most impressive performance in terms of synthetic accessibility (SAS), demonstrating the practical applicability of the model for molecule generation.

Overall, NIMO achieved a major breakthrough as a fragment-based model compared to the FBMG and is in no way inferior to the state-of-the-art SMILES-based model. Indeed, novelty remains a future endeavor in the field of molecular generation. Next, three practical molecular generation tasks, including structure-oriented (terpenoids), bioactivity-oriented (antimalarials), and target-driven (antibacterial) tasks, were performed to demonstrate the applicability of our model.

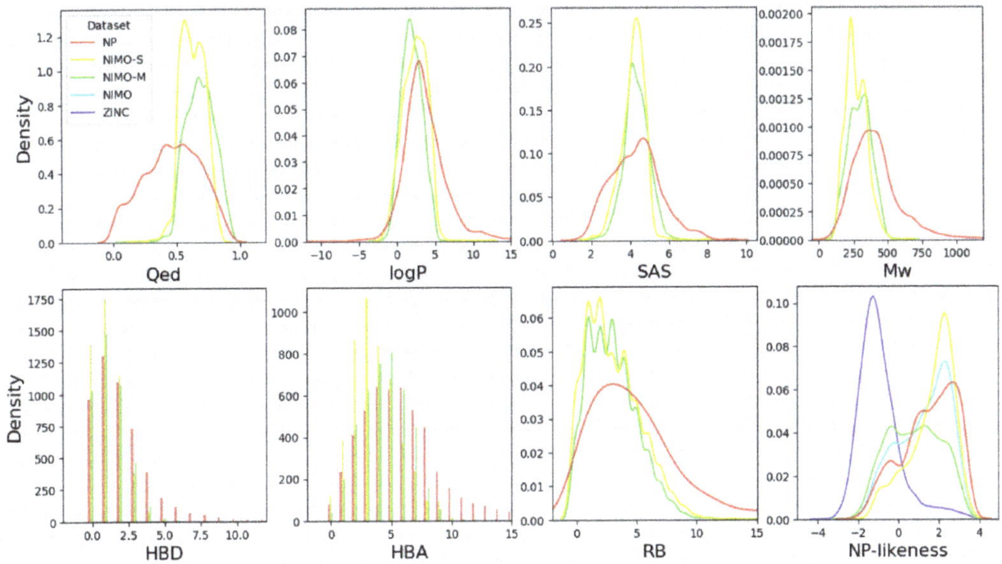

Figure 2. Property distributions of the molecules generated by NIMO. NP (red) refers to the train set, while the ZINC (purple) dataset is identified as synthetic molecules. NIMO (blue) is from the integration of molecules generated by NIMO-M (green) and NIMO-S (yellow). QED is "quantitative estimate of drug likeness"; logP indicates "octanol/water partition coefficient"; SAS indicates "synthetic accessibility score"; Mw indicates "molecular weight"; HBD indicates "hydrogen bond donor"; HBA indicates "hydrogen bond acceptor"; RB indicates "rotatable bond".

Table 1. The conditional and MOSES evaluation metrics for the generated molecules. The detailed definitions are provided in Supplementary Materials.

	Models	NIMO-M	NIMO-S	MCMG	QBMG	FBMG
Conditional metrics	Validity	94.5%	**99.3%**	95.0%	94.5%	42.9%
	Uniqueness	99.7%	99.1%	98.4%	**99.9%**	98.5%
	Novelty	61.0%	**77.8%**	65.7%	42.2%	99.9%
MOSES metrics	FCD↓ [a]	**3.71**	11.2	4.52	19.2	6.11
	SNN↓	0.87	**0.65**	0.71	0.95	0.51
	Frag↓	0.85	**0.77**	0.95	0.99	0.48
	Scaf↓	0.67	0.83	**0.65**	0.66	0.57
	IntDiv	**88.3%**	86.5%	87.8%	86.6%	73.9%
	Novelty	71.4%	**89.0%**	79.5%	52.4%	99.9%
	SAS↓	**0.78**	0.91	1.22	0.87	0.94

[a] ↓ The lower, the better. FCD refers to "Fréchet ChemNet Distance", which is a metric to predict biological activities based on a deep neural network; SNN refers to "nearest neighbor similarity"; Frag/Scaf refers to "fragment/scaffold similarity"; IntDiv refers to "internal diversity". Bold text indicates the best result.

2.2. Terpenoid Generation

In order to evaluate whether the model can generate molecules targeting specific structure regions, we specifically designed the generation task of anchoring complex polycyclic terpenoids, which is the biggest class of NPs. As shown in Table 2, four models were trained based on the TeroKIT database. Herein, NIMO-S′ added ring separation and ring recombination (edge fusion) functions to NIMO-S to compensate for the lack of scaffold diversity. The additional step was helpful for reducing the size and weight of the motifs, as shown in Figure S3. Subsequently, all generated molecules were judged in terms of whether the structure belonged to terpenoids by NPClassifier [36]. As summarized in Table 2, NIMO-S′ had the best performance (95.4%) for effectively constructing terpenoid compounds, followed by NIMO-S, QBMG, and QCMG. This indicates that our NIMO models can generate more norm-compliant molecules with the constraints of the established structure rules.

Table 2. The performance for terpenoid generation. "Success" means the success rate of molecules predicted as terpenoids by NPClassifier; "coverage" represents the proportion of the number of unique RSs/FGs extracted from the generated set and existing in the training set to the total RSs/FGs of the generated set; "recovery" represents the proportion of the number of unique RSs/FGs extracted from the generated set and existing in the training set to the unique RSs/FGs of the generated set.

	Metrics	NIMO-S	NIMO-S′	MCMG	QBMG
Terpenoids	Success	91.9%	**95.4%**	71.2%	89.7%
Ring systems	Coverage	27.5%	**29.8%**	28.1%	8.3%
(RSs) [a]	Recovery	**99.4%**	69.5%	62.4%	10.6%
Functional groups	Coverage	5.9%	**6.2%**	4.3%	4.9%
(FGs) [b]	Recovery	**93.2%**	89.7%	58.1%	47.1%

[a,b] RSs and FGs were automatically extracted based on RDKit in an unbiased way. Bold text indicates the best result.

The functional groups (FGs) and ring systems (RSs) were then identified for the 5000 generated molecules and the TeroKit dataset [37,38]. As shown in Table 2, NIMO-S and NIMO-S′ exhibited good coverage of scaffolds present in the training set, according to "coverage" and "recovery". Obviously, our scaffold-based NIMO model can maximally reproduce the substructural features of the original training set. In addition, a growing body of evidence supports the effectiveness of retaining specific substructures (e.g., core scaffolds) or general structural features (e.g., RSs and FGs) for inheriting the biological relevance of natural products [39,40]. This is of greater importance for the structural modification of proven scaffolds in drug screening. Therefore, we further developed a scaffold-based scenario for a more elaborative evaluation.

Two motifs were chosen as core seeds and utilized for molecular generation (Figure 3). It was found that NIMO-S could reproduce the same modifications at conserved sites of the core scaffold. Moreover, due to the correct definition of the extension sites, it offered diverse modifications for the scaffold with different functional groups. For example, the same functional group modifications, such as methyl and hydroxyl groups, were present in the generated molecules at some derivation sites (e.g., C-4) compared to the real molecules seen from scaffold 1. More importantly, NIMO-S decorated various substructures at extension sites to generate diverse derivatives, such as the long side chain at C-11 of scaffold 1 (blue) and C-21/16 of scaffold 2 (green). More structural analysis of the generated molecules is provided in Figure S4, and Figure S5 depicts that NIMO was also able to reconstruct a similar chemical space of terpenoids.

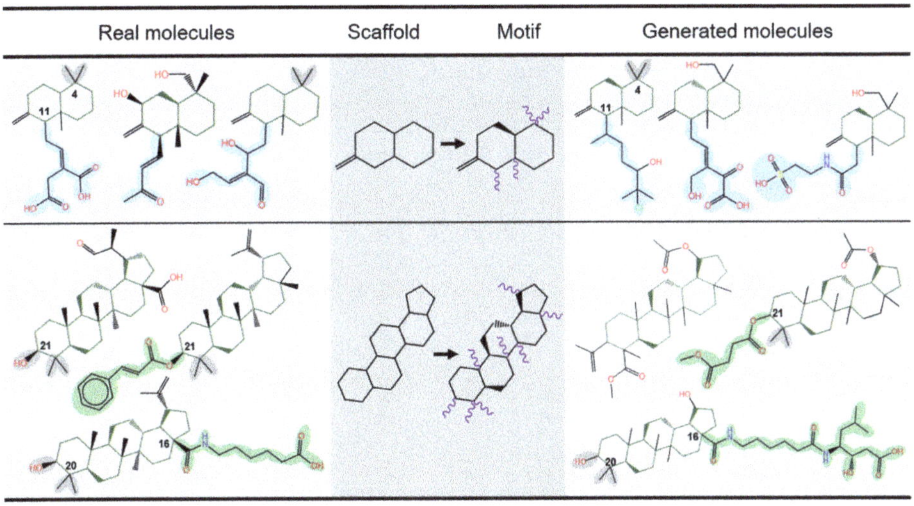

Figure 3. The typical cases of derivatives generated by NIMO-S in a scaffold-based scenario. The middle columns are motifs corresponding to the two randomly selected scaffolds from the terpenoid training set. The left and right columns are real molecules containing the structure of scaffolds and molecules generated by NIMO-S, respectively. The extension sites are highlighted in purple.

2.3. Antimalarial Activity-Oriented Molecular Generation

As NPs serve as the major source of lead compounds against malaria [41,42], the motif-based NIMO model was used to discover potent new antimalarials. Large-scale predictions of potential antimalarial compounds were made on MAIP [43]. Valid molecules were sampled from models under evaluation setting 3 (see Section 4.5). As a result, the histogram distribution of the predicted scores from MCMG roughly followed a normal distribution, like the training data, whereas those from NIMO-M appeared unsmoothed and discontinuous (see details in Figure S6). The antimalarial activity prediction of molecules generated by NIMO-M and MCMG are summarized in the left four columns of Table 3. NIMO-M excelled in two of the three enrichment factor metrics and outperformed MCMG, with an overall activity rate of 55.9%, approximately 46% higher than the training dataset.

Table 3. Summary of anti-malarial activity-oriented molecular generation.

	Train	NIMO-M	MCMG	NIMO-M′
Samples	744,986	5000	5000	1000
EF [50%] [a]	20.07	46.82	44.99	68.22
EF [10%]	44.36	72.09	69.11	81.33
EF [1%]	80.4	81.97	89.17	92.21
Active %	10.0%	55.9%	52.1%	85.5%

[a] EF means enrichment factor provided by MAIP. EF [X] is the hit rate (the proportion of active compounds) within a defined sorted fraction divided by the total hit rate.

The first quartile, indicating the novel molecules generated by two models, was quantitatively close, as shown in Figure 4a. NIMO showed dense enrichment at high similarity around the third quartile. Four high-frequency motifs in the top 10% of active molecules generated by NIMO-M are listed in Figure 4b. Molecules containing high-frequency motifs yielded a more dominant predicted score, as shown in Figure 4c. This reflects that the NIMO models were capable of sampling active motifs, which made up a large part of the total sampling volume. In particular, Figure 4d shows that Motif2 had more potential for exploring antimalarial activity. Thus, Motif2 was seeded into the trained NIMO-M for resampling, which was named NIMO-M′. As we expected, NIMO-M′ achieved a boost in the enrichment ability toward anti-malarial activity. As shown in the rightmost column of Table 3, three enrichment factors were significantly increased, with the overall activity ratio raised by approximately 75.5% over the training data. The result also shows that the predicted activity was more susceptible to fragments than tokens tokenized by SMILES. To facilitate data analysis, we also visualized the chemical spatial distribution of the training set and generated molecules using TMAP [44], as shown in Figure S7. If we regard the location of the NIMO-M′-generated molecules as the highly active region (orange dots), then we can observe that the NIMO-M-generated molecules were clustered nearby, resulting in a high molecular density. This illustrates that NIMO-M exhibited a structural preference over the active region formed by the dominant motifs, which is advantageous for realistic molecular generation practices.

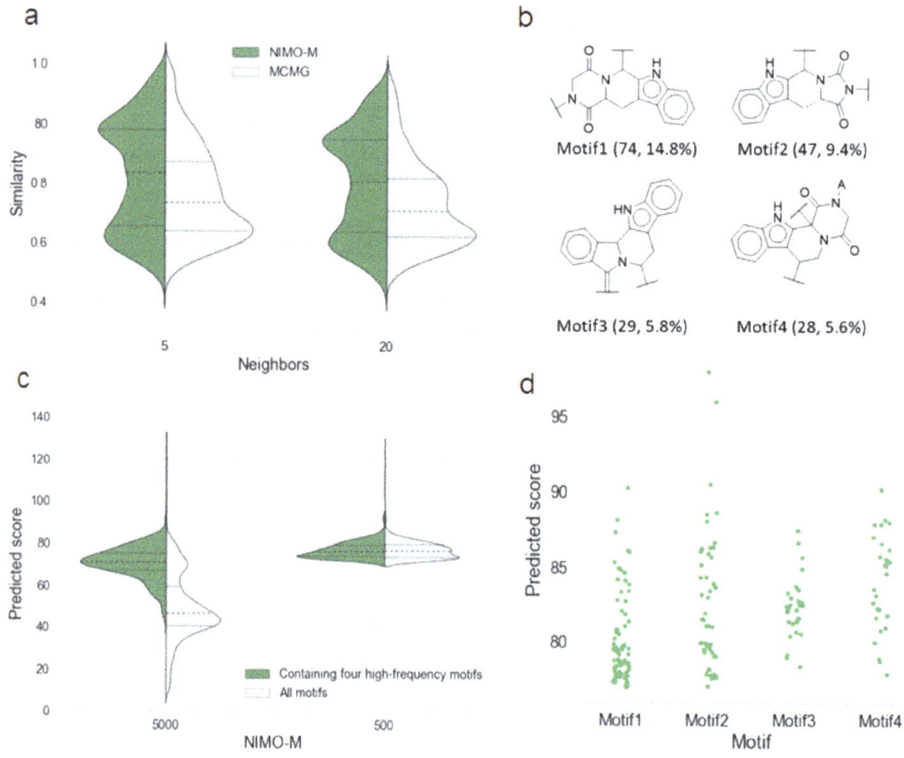

Figure 4. (**a**) Distribution of the average Tanimoto similarity between fingerprints of molecules generated by models (NIMO-M, MCMG) and the nearest neighbor molecules from the training set. (**b**) Four high-frequency motifs in the top 500 molecules generated by NIMO-M ranked by anti-malarial activity score, along with the number and percentage of molecules categorized by the motifs in brackets. (**c**) Violin plots of predicted scores for molecules generated by NIMO-M. The left represents the 5000 molecules, and the right represents the top 500 molecules scored by predicted scores. (**d**) Scatter plots of predicted scores of molecules categorized by the above four motifs.

2.4. Pocket-Based Molecular Generation

NIMO showed excellent enrichment ability in the above specific activity-oriented task. Next, we spotlighted the fragment-derived methods and strategies for the effectiveness of virtual library development [45,46]. We proposed a general approach to design antibacterial discovery libraries. Briefly, we collected an antibacterial dataset against experiment-relevant Gram-positive and Gram-negative bacteria, covering fifteen common bacterial species.

Then, a phenotypic antimicrobial model based on a kNN classifier against the bacteria panel was constructed by molecular fingerprinting analysis. Then, the classifier was used to further predict targeted bacterial species for molecules generated by NIMO-M. Finally, we obtained a target annotated intelligent library. See Figure S8 for detailed steps.

Here, the Lactobacillus-targeted compounds from the above antibacterial library were selected for further analysis. The compounds falling under the targets (2HMG, 1BO7) were docked into the associated protein pockets, as shown in Table 4. According to docking results by MOE [47], among the 5000 molecules generated for 2HMG, 82 candidates were predicted by molecular fingerprinting analysis, and the docking scores of 26 molecules were lower than those of the native ligand (the lower score the better). For the top 1000 molecules, 10 of the 15 candidates had a dominant docking score. There was also a significant proportion of compounds for 1B07. Moreover, compounds with RMSD values of less than 2 Å to native ligands indicate that the original binding poses could be well recovered by NIMO.

Table 4. MOE docking result of compounds generated by NIMO-M.

PDB	2HMG (CHEMBL2902)		1BO7 (CHEMBL5328)	
Compounds	**1000**	**5000**	**1000**	**5000**
Predicted candidates	15	82	93	294
Docking score < native	10	26	10	23
RMSD < 2	10	65	48	104

Figure 5 showed four high-quality binding poses of compounds against Lactobacillus selected from the above virtual antibacterial library. There were favorable MOE docking poses after overlay with native ligands in three protein pockets. Besides a high 3D shape similarity, they had a better docking score compared to the co-crystal ligand, indicating a positive binding affinity with the pocket. Meanwhile, three indicators (QED, SAS, SI) also showed that NIMO could deliver chemically reasonable compounds. On the other hand, compounds 3 and 4 appeared to have the same topological structures of 2,4-diaminopyrimidine rings (in a circle with dashed lines) with native ligands [48,49]. This suggests that NIMO can both reproduce the key pharmacophore features of the active ligand and capture more attractive fragments from the training set.

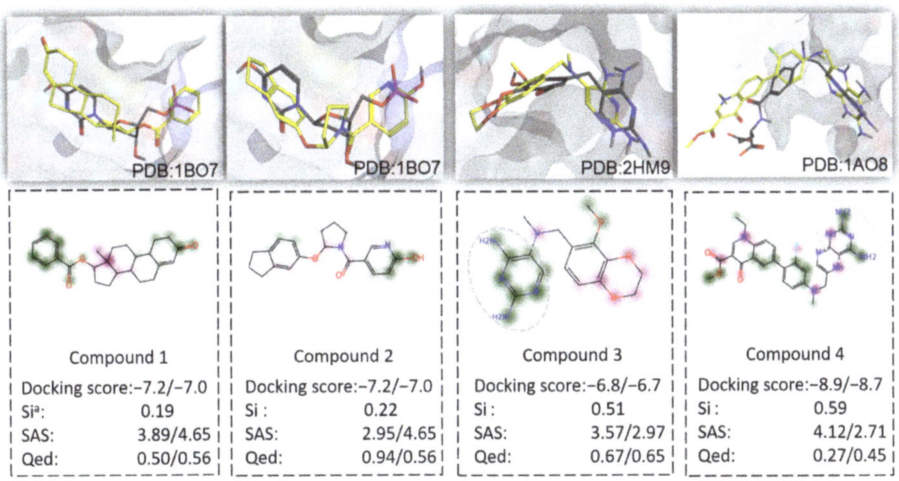

Figure 5. Examples of reasonable compounds generated by NIMO-M. [a] SI represents similarity index. The data behind the "/" is the value for native ligands.

2.5. Discussion

NIMO is a fragment-based generator capable of handling stereochemical information from natural products. In the case of NIMO-M, the attachment points that tagged the cleaved bonds retained chiral information in the initial fragments after fragment extraction. Moreover, our model also accounted for constraint structural optimization and allowed derivative compounds to be formed starting from a specified substructure, which was extremely useful in practice. In addition to the above innovations to address basic challenges, the most notable differences were in the performance for NIMO in comparison to other methods. (1) NIMO can generate more norm-compliant structural categories under the intended structural disciplines. For example, our model generated terpenoid structures with a success rates of 95.4% (NIMO-S') and 91.9% (NIMO-S), which outperformed the baseline models in the structure-based generation task (Table 2). (2) The molecules generated by NIMO can inherit the biological relevance in a friendly way by maximizing the reproduction of substructures found in natural products. For example, our model reproduced ring systems and functional groups that were pre-existing in the training dataset, surpassing baseline models in terms of coverage and recovery metrics (Table 2). (3) NIMO can detect potentially privileged motifs that contribute to activity and enrich more active molecules as a result (Figure 4). Meanwhile, NIMO showed a strong structural preference for highly active regions rather than a uniform distribution (Figure S7). (4) In terms of a granularly fragment-based algorithm, the high efficiency of fragment extraction brings smaller motifs and consequently increases the molecular diversity. This was confirmed in the comparison result of motif extraction among fragment-based models (Figure S1, Table S1). For instance, the mean weight of the motifs produced by our model was 218.5 g/mol (NIMO-M) and 238.2 g/mol (NIMO-S), significantly lower than that of the same fragment-based model, FBMG, which stood at 407.2 g/mol. On the other side, the high reconstruction accuracy (99.9%) warrants that only the correct motif sequences were fed into the model; thus, it circumvented the puzzles of where to attach the new fragment and which chemical bond to choose, as required by conventional fragment-based models. Nevertheless, the future development of NIMO still comes up against a few open questions. NIMO can mimic fragment rearrangement, ring separation, and ring combination (edge fusion), but some other more complicated design strategies, such as opening/closing ring and bridged ring, were not realized to generate pseudo-NPs in the current work [11,50,51].

3. Methods

3.1. Data Preparation

In this work, all available datasets were collected from public domains, including COCONUT [52], TeroKIT [53,54], ChEMBL [55], and the study of Andreas Verras et al. [56], as detailed in Section 4.1. The data were filtered according to molecule standardization for consistency. The complete procedure consisted of desalination, charge neutralization, removal of glycosylation, and checking of molecular validity. Also, duplicates were removed. As a result, all natural products with stereochemistry were collected in the form of canonical SMILES strings. The filter was followed by calculating each molecular entry with molecular descriptors. They were used as constraints for the training model and as metrics for the model evaluation.

3.2. Motif Sequence Generation

We utilized two tailor-made methods to generate motif sequences for NPs, applied to the motif-based model (NIMO-M) and the scaffold-based model (NIMO-S), respectively. First, we defined a motif S_i as a subgraph of molecule G. Second, we decomposed molecule G into fragments by breaking bonds specifically by the fragmentation rules. Many rules were included, but not limited to the BRICS [57] and Murcko [58] fragmentation methods (see Section 4.2). The generated fragments contained some dummy atoms with their original bond IDs, allowing the original connection to be memorized. See Supplementary for details of the fragmentation protocol. Next, the generated fragment sequence was canonicalized according to the dummy atom IDs of the initial fragments. The sequence order was generated in such a way that the original molecule could be reconstructed without using the dummy atom IDs. We denoted each canonicalized fragment with rich semantic information as a motif S_i, and the motif served as the basic unit for training the model. The procedure for canonicalizing fragmented sequences is provided in Table S2, and two cases of motif sequence generation are presented in Figures S9 and S10.

3.3. Molecular Reconstruction Verification

Molecular reconstruction verification was applied after the canonical motif sequences were generated. The verification was used to filter out a small number of invalid sequences, as illustrated in Figure S11. An example of molecular reconstruction is outlined in Figure S12. A rigorous examination of all motif sequences ensured a molecular reconstruction accuracy of 99.9%, regardless of chirality differences. In contrast, the HierVAE-decoder [21] also utilized motifs as building blocks for generation, though it only reached an accuracy of around 80%. This indicates that the reserved dummy atoms in the motif allowed us to omit the process of attachment prediction and reduce the loss. This significant boost gave us great confidence in the reliability and interpretability of the fragment-based model.

3.4. Model Architecture

As shown in Figure 6, the core of NIMO used a conditional transformer architecture to generate NP-derived molecules with desirable properties. First, each pre-processed input sequence, including constraints, motif information, and motif sequence, was viewed as a sentence and a vocabulary was constructed. "Motif info" represents the number of attachment points automatically extracted from the motif sequence in advance. Second, the sentence was fed into input embedding, followed by the addition of positional encoding. Here, the standard sinusoidal positional encoding allowed the transformer to preserve the relative position of words in a sentence. The core architecture consisted of multiple decoder stacks. Each decoder layer had a multi-head self-attention sub-layer and a position-wise feedforward network (FFN) sub-layer. The masked multi-head self-attention layer ensured that the prediction of the current position relied only on the sequence embedding information prior to that position. The self-attention layer applied scaled dot-product attention functions and facilitated the model to capture information from different subspaces at different positions. The formula of the attentional mechanism can be described according to the following equation:

$$Attention(Q, K, V) = softmax(\frac{QK^T}{\sqrt{d_k}})V \quad (1)$$

This formula required the introduction of the query (Q), a key (K), and a scaling factor d_k. Then, a softmax function was used to obtain the weights of the values (V). The FFN adopted two layers of fully connected layers. Next, ReLU was an activation function followed by a layer normalization procedure. Then, a residual connection was applied to ease the gradient disappearance and allow for a deeper network. The decoder outputs yielded a probability distribution over all latent semantic rules for each time step. The input sequence was expressed as $X = x_1, \ldots, x_k$. Since the model inputs contained desirable properties (as a constraint condition c), the model was trained to minimize the following negative log-likelihood:

$$L(x) = -\sum_{i=1}^{k} log\ p\ (x_i | x_0, \ldots, x_{i-1}, c) \quad (2)$$

During model sampling, linear and softmax layers produced an output probability for the next word according to a learnt conditional probability distribution:

$$x_i \sim p(x_0, \ldots, x_{i-1}, c) \quad (3)$$

The sampling problem was defined as the search for the most probable hypothesis y* according to a trained model and a set of constraints c. If V was the search space formed by vocabulary combinations, y^* was calculated by the following equation:

$$y^* = \text{argmax}\ p(y|x,c) \qquad y \in V \tag{4}$$

The decoder recursively generated the subsequent samples by adopting a beam search algorithm [59]. The beam search kept the K locally highest probability candidates at each time step t, where the hyperparameter K was referred to as the beam width. The recursion was performed until all sampling sequences ended in the character "EOS" or the predefined maximum time step T was reached. The maximum search space of this algorithm in one generation process was related to the spatial complexity O(TKV). More details on the sensitivity analysis of parameter K can be found in Figure S13. Herein, we modulated the probability distribution in some scenarios by avoiding the occurrence of unreasonable fragments and preventing the beam from going in a repetitive direction. Next, top N hypotheses y^* were selected from the searched set according to scoring accumulated probabilities. Finally, a molecular reconstruction algorithm transformed the N sequences to the final molecular structure, which was a reverse procedure of the molecular fragmentation and canonicalization.

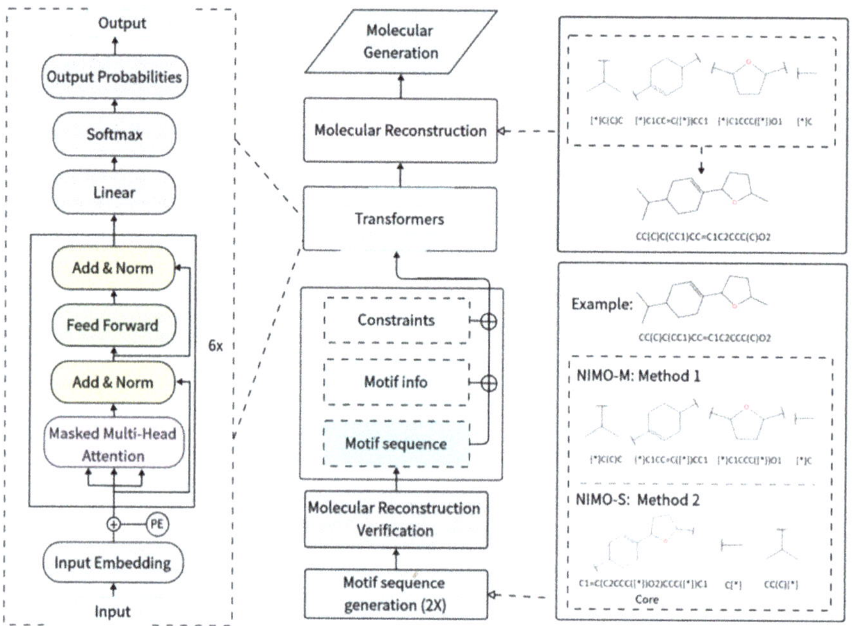

Figure 6. The workflow of NIMO. Middle: The model-training preparation process includes motif sequence canonicalization and molecular reconstruction verification, which results in unique canonical motif sequences. Then, each input sequence includes constraints, motif information, and a motif sequence, which are used to train a conditional transformer architecture. During the model sampling phase, the sampled motifs undergo molecular reconstruction to form the final complete molecules. Left: The architecture of the conditional transformer neural network. Right: the examples of molecular reconstruction and motif sequence generation (including two methods, corresponding to NIMO-M and NIMO-S). "*" denotes a dummy atom within the motif.

4. Experiment Configuration

4.1. Datasets

4.1.1. COCONUT

The natural product structures in the training dataset were downloaded from the collection of open natural products (COCONUT, https://coconut.naturalproducts.net/ (accessed on 1 May 2023)) in absolute SMILES (includes stereochemical information) format, which initially contained about 744,986 unique canonical SMILESs.

4.1.2. TeroKIT

The terpene dataset TeroKIT was obtained from our group's previous study. About 173,914 annotated terpenoids were collected. More details can be accessed at (accessed on 1 May 2023).

4.1.3. Anti-Malarial Experimental Activity Data Set

The anti-malarial experimental activity dataset was MMV-St. Jude, which was obtained from the study reported by Andreas Verraset al. It contains 2507 positive compounds.

4.1.4. Antibacterial Dataset

All antibacterial compounds against common Gram-positive and Gram-negative bacteria were retrieved from the ChEMBL dataset (255,788).

4.2. Fragment Extraction

Given a compound library, we used two special fragmentation methods to transform the molecular graph into a sequence of fragments, which were applied to NIMO-M and NIMO-S. First, we defined a motif $S_i = (V_i, E_i)$ as a subgraph of molecule G, where V_i is the set of atoms (vertices) and E_i is the set of bonds (edges). To extract motifs, we decomposed molecule G into fragments by breaking bonds specified by the following rules. In NIMO-M, (1) find all the single bonds (μ, ν) ∈ E, where u is in a ring, and ν is in an off-ring or is in another ring. Bonds (μ, ν) are undirected. (2) Find all the bonds that meet BRICS [57]. In NIMO-S, (1) find all the bonds between the Murcko scaffold and the side chains. (2) Find a bond (u, v) in Murcko [58] scaffold that represents a shared edge in a fused ring. Meanwhile, the bond (μ, v) divided the Murcko scaffold S_1 future into two subgraphs (S_2, S_3). The fragment extraction resulted in initial fragments containing dummy atoms. Therefore, we obtained motifs such as "C1CC[*][*]C1" and "[*]1 = [*]C = CC1", where the atom types u and v were further replaced by dummy atom [*].

4.3. Evaluation Setting 1

The training data were from the COCONUT dataset. A total of 5000 molecules were sampled from the multi-constraint models. Specifically speaking, three of these molecular features were selected as constraints to train the models in our work. The QED, logP, and SAS were expressed as scalars. Each molecule was labeled with different attributes based on a customized threshold value, such as "good logP". These labels were applied to train the biased model as constraint codes. Model training and optimization of hyperparameters are provided in Table S3. Finally, we plotted the distributions by the statistics and analysis of partial descriptors.

4.4. Evaluation Setting 2

FBMG was a fragment-based generative model as well. QBMG was a natural product-focused SMILES-based generative model. MCMG was one of the most advanced SMILES-based generative models and was also used for contrast. It should be pointed out that MCMG in this article specifically refers to MCMGM, where distilled molecules (DM) were taken as the knowledge distillation method. NIMO was trained just like evaluation setting 1.

4.5. Evaluation Setting 3

First, predictions of potential malaria-inhibiting compounds from the COCONUT dataset were made on MAIP. The ultimate predicted output of MAIP is a model score. Here, we defined 44.36 as a score threshold. This meant that 10% of natural products with a model score > 44.36 were labeled as anti-malaria active data, while the remaining 90% were labeled as inactive. Moreover, a portion of the anti-malarial experimental activity dataset (2507) was coupled with labeled compounds to train the models. Finally, 5000 valid molecules were sampled from models conditioned on a given activity constraint. Only MCMG was allowed as the baseline model to perform activity-constraint molecular generation.

4.6. Baseline Models

4.6.1. MCMG

We downloaded the code from the official repository https://github.com/jkwang93/MCMG (accessed on 1 July 2023). The MCMG as a SMILES-based model was unable to handle stereo information. There were 316,864 unique "flat" (with no stereochemistry) NPs in the training set after the elimination of stereo information and de-duplication. Additionally, MCMG was slightly modified to carry out its training process with the same constraints as ours (QED, logP, SAS). Then, the model was trained according to the process described in the original study.

4.6.2. QBMG

We downloaded the code from the official repository https://github.com/SYSU-RCDD/QBMG (accessed on 1 July 2023). QBMG was able to handle stereo information as a quasi-biogenic molecule generator. QBMG was trained without constraints in order to avoid major human intervention.

4.6.3. FBMG

We downloaded the code from the official repository https://github.com/marcopodda/fragment-based-dgm (accessed on 1 July 2023). FBMG was trained without constraints in order to avoid major human intervention and maintain the original function. The model cannot generate molecules with stereo information. Default parameters were used to train the model.

4.7. NP-Likeness Score

The NP-likeness score was calculated using RDKit-based implementation of the method described in the original article, which can be found in the repository https://github.com/rdkit/rdkit/tree/master/Contrib/NP_Score (accessed on 1 March 2023).

4.8. NPClassifier

The NPClassifier is a deep learning-based automated structural classification of NPs. Herein, the final statistic in Table 2 depends on the terpenoids classified by the NPClassifier. The detailed implementation can be found by visiting this file: https://pubs.acs.org/doi/suppl/10.1021/acs.jnatprod.1c00399/suppl_file/np1c00399_si_003.pdf (accessed on 1 July 2023).

4.9. MAIP

The malaria inhibitor prediction (MAIP) is accessible through https://www.ebi.ac.uk/chembl/maip/ (accessed on 1 July 2023). When using the web service to predict blood-stage malaria inhibitors, MAIP returns a predicted model score. A higher score means greater enrichment.

5. Conclusions

In this work, we proposed a new design strategy (named NIMO) for molecule generation to efficiently explore the vast chemical space of natural products. NIMO is helpful for discovering bioactive NP-like compounds and structural modification of NPs. Two sets of motif extraction methods were used to fragment molecule structures and derive motifs in semantically meaningful sequences. A constrained transformer framework was developed to capture rich semantic information and implicit linking rules. As a result, NIMO demonstrated superior performance across three typical applications (structure-guided, activity-oriented, and pocket-based). Although there is still room for further improvements, we believe that NIMO could provide a general computational framework for fragment-to-lead design to accelerate the construction of high-quality pseudo-natural product libraries. This approach can be applied to various scenarios, such as multi-objective structural optimization, scaffold-based lead optimization, and activity-oriented enrichment based on dominant fragments, thereby facilitating drug discovery for natural products.

Supplementary Materials: The following supporting information can be downloaded at: https://www.mdpi.com/article/10.3390/molecules29081867/s1, ref. [60] is cited in the Supplementary Materials.

Author Contributions: X.S. designed and implemented the deep learning model, performed the model training, and analyzed the data. R.W. and J.L. designed and supervised the project. X.S., T.Z. and N.C. discussed the results. All authors revised the article critically for important intellectual content. All authors have read and agreed to the published version of the manuscript.

Funding: This work was supported by the National Key Research and Development Program of China (2023YFC3404900) and the Key Area Research and Development Program of Guangdong Province (2022B1111080005). We are also thankful for the Top-Notch Young Talents Program of China.

Institutional Review Board Statement: Not applicable.

Informed Consent Statement: Not applicable.

Data Availability Statement: The code of NIMO is publicly available from the GitHub repository: https://github.com/shenxj9/NIMO (accessed on 22 March 2024).

Conflicts of Interest: Author Jiabo Li was employed by the company ChemXAI Inc. The remaining authors declare that the research was conducted in the absence of any commercial or financial relationships that could be construed as a potential conflict of interest.

References

1. Grigalunas, M.; Brakmann, S.; Waldmann, H. Chemical Evolution of Natural Product Structure. *J. Am. Chem. Soc.* **2022**, *144*, 3314–3329. [CrossRef] [PubMed]
2. Atanasov, A.G.; Zotchev, S.B.; Dirsch, V.M.; Supuran, C.T.; International Natural Product Sciences Taskforce. Natural products in drug discovery: Advances and opportunities. *Nat. Rev. Drug Discov.* **2021**, *20*, 200–216. [CrossRef] [PubMed]
3. Newman, D.J.; Cragg, G.M. Natural Products as Sources of New Drugs over the Nearly Four Decades from 01/1981 to 09/2019. *J. Nat. Prod.* **2020**, *83*, 770–803. [CrossRef] [PubMed]
4. Rodrigues, T.; Reker, D.; Schneider, P.; Schneider, G. Counting on natural products for drug design. *Nat. Chem.* **2016**, *8*, 531–541. [CrossRef] [PubMed]
5. Ding, Y.; Xue, X. Medicinal Chemistry Strategies for the Modification of Bioactive Natural Products. *Molecules* **2024**, *29*, 689. [CrossRef] [PubMed]
6. Chavez-Hernandez, A.L.; Sanchez-Cruz, N.; Medina-Franco, J.L. A Fragment Library of Natural Products and its Comparative Chemoinformatic Characterization. *Mol. Inform.* **2020**, *39*, e2000050. [CrossRef] [PubMed]
7. Wetzel, S.; Bon, R.S.; Kumar, K.; Waldmann, H. Biology-oriented synthesis. *Angew. Chem. Int. Ed. Engl.* **2011**, *50*, 10800–10826. [CrossRef] [PubMed]
8. Gagare, S.; Patil, P.; Jain, A. Natural product-inspired strategies towards the discovery of novel bioactive molecules. *Future J. Pharm. Sci.* **2024**, *10*, 55. [CrossRef]
9. Karageorgis, G.; Foley, D.J.; Laraia, L.; Waldmann, H. Principle and design of pseudo-natural products. *Nat. Chem.* **2020**, *12*, 227–235. [CrossRef]
10. Bag, S.; Liu, J.; Patil, S.; Bonowski, J.; Koska, S.; Schölermann, B.; Zhang, R.; Wang, L.; Pahl, A.; Sievers, S. A divergent intermediate strategy yields biologically diverse pseudo-natural products. *Nat. Chem.* **2024**, 1–14. [CrossRef]
11. Nelson, A.; Karageorgis, G. Natural product-informed exploration of chemical space to enable bioactive molecular discovery. *RSC Med. Chem.* **2021**, *12*, 353–362. [CrossRef] [PubMed]
12. Hou, S.H.; Zhou, F.F.; Sun, Y.H.; Li, Q.Z. Deconstructive and Divergent Synthesis of Bioactive Natural Products. *Molecules* **2023**, *28*, 6193. [CrossRef] [PubMed]
13. Lehn, J.M. Dynamic combinatorial chemistry and virtual combinatorial libraries. In *Essays in Contemporary Chemistry: From Molecular Structure towards Biology*; Wiley Online Library: Hoboken, NJ, USA, 2001; pp. 307–326.

14. Cheng, Y.; Gong, Y.; Liu, Y.; Song, B.; Zou, Q. Molecular design in drug discovery: A comprehensive review of deep generative models. *Brief. Bioinform.* **2021**, *22*, bbab344. [CrossRef] [PubMed]
15. Fromer, J.C.; Coley, C.W. Computer-aided multi-objective optimization in small molecule discovery. *Patterns* **2023**, *4*, 100678. [CrossRef] [PubMed]
16. Born, J.; Manica, M. Regression transformer enables concurrent sequence regression and generation for molecular language modelling. *Nat. Mach. Intell.* **2023**, *5*, 432–444. [CrossRef]
17. Wang, J.; Hsieh, C.-Y.; Wang, M.; Wang, X.; Wu, Z.; Jiang, D.; Liao, B.; Zhang, X.; Yang, B.; He, Q.; et al. Multi-constraint molecular generation based on conditional transformer, knowledge distillation and reinforcement learning. *Nat. Mach. Intell.* **2021**, *3*, 914–922. [CrossRef]
18. Vaswani, A.; Shazeer, N.; Parmar, N.; Uszkoreit, J.; Jones, L.; Gomez, A.N.; Kaiser, L.; Polosukhin, I. Attention is all you need. *Adv. Neural Inf. Proc. Syst.* **2017**, *30*, 5998–6008.
19. Zheng, S.; Yan, X.; Gu, Q.; Yang, Y.; Du, Y.; Lu, Y.; Xu, J. QBMG: Quasi-biogenic molecule generator with deep recurrent neural network. *J. Cheminform.* **2019**, *11*, 5. [CrossRef]
20. Yoshikai, Y.; Mizuno, T.; Nemoto, S.; Kusuhara, H. Difficulty in chirality recognition for Transformer architectures learning chemical structures from string representations. *Nat. Commun.* **2024**, *15*, 1197. [CrossRef]
21. Jin, W.; Barzilay, R.; Jaakkola, T. Hierarchical generation of molecular graphs using structural motifs. In Proceedings of the International Conference on Machine Learning, Virtual, 13–18 July 2020; pp. 4839–4848.
22. Podda, M.; Bacciu, D.; Micheli, A. A Deep Generative Model for Fragment-Based Molecule Generation. *Int. Conf. Artif. Intell. Stat.* **2020**, *108*, 2240–2250.
23. Ortholand, J.Y.; Ganesan, A. Natural products and combinatorial chemistry: Back to the future. *Curr. Opin. Chem. Biol.* **2004**, *8*, 271–280. [CrossRef] [PubMed]
24. Harvey, A.L.; Clark, R.L.; Mackay, S.P.; Johnston, B.F. Current strategies for drug discovery through natural products. *Expert Opin. Drug Discov.* **2010**, *5*, 559–568. [CrossRef] [PubMed]
25. Davison, E.K.; Brimble, M.A. Natural product derived privileged scaffolds in drug discovery. *Curr. Opin. Chem. Biol.* **2019**, *52*, 1–8. [CrossRef] [PubMed]
26. Mullowney, M.W.; Duncan, K.R.; Elsayed, S.S.; Garg, N.; van der Hooft, J.J.J.; Martin, N.I.; Meijer, D.; Terlouw, B.R.; Biermann, F.; Blin, K.; et al. Artificial intelligence for natural product drug discovery. *Nat. Rev. Drug Discov.* **2023**, *22*, 895–916. [CrossRef] [PubMed]
27. Meyers, J.; Fabian, B.; Brown, N. De novo molecular design and generative models. *Drug Discov. Today* **2021**, *26*, 2707–2715. [CrossRef]
28. Jinsong, S.; Qifeng, J.; Xing, C.; Hao, Y.; Wang, L. Molecular fragmentation as a crucial step in the AI-based drug development pathway. *Commun. Chem.* **2024**, *7*, 20. [CrossRef] [PubMed]
29. Cheng, A.H.; Cai, A.; Miret, S.; Malkomes, G.; Phielipp, M.; Aspuru-Guzik, A. Group SELFIES: A robust fragment-based molecular string representation. *Digit. Discov.* **2023**, *2*, 748–758. [CrossRef]
30. Lim, J.; Hwang, S.-Y.; Moon, S.; Kim, S.; Kim, W.Y. Scaffold-based molecular design with a graph generative model. *Chem. Sci.* **2020**, *11*, 1153–1164. [CrossRef] [PubMed]
31. Tan, X.; Li, C.; Yang, R.; Zhao, S.; Li, F.; Li, X.; Chen, L.; Wan, X.; Liu, X.; Yang, T. Discovery of pyrazolo [3, 4-d] pyridazinone derivatives as selective DDR1 inhibitors via deep learning based design, synthesis, and biological evaluation. *J. Med. Chem.* **2021**, *65*, 103–119. [CrossRef]
32. Seidel, T.; Wieder, O.; Garon, A.; Langer, T. Applications of the Pharmacophore Concept in Natural Product inspired Drug Design. *Mol. Inform.* **2020**, *39*, e2000059. [CrossRef]
33. Ertl, P.; Roggo, S.; Schuffenhauer, A. Natural product-likeness score and its application for prioritization of compound libraries. *J. Chem. Inf. Model.* **2008**, *48*, 68–74. [CrossRef] [PubMed]
34. Sterling, T.; Irwin, J.J. ZINC 15—Ligand Discovery for Everyone. *J. Chem. Inf. Model.* **2015**, *55*, 2324–2337. [CrossRef] [PubMed]
35. Polykovskiy, D.; Zhebrak, A.; Sanchez-Lengeling, B.; Golovanov, S.; Tatanov, O.; Belyaev, S.; Kurbanov, R.; Artamonov, A.; Aladinskiy, V.; Veselov, M.; et al. Molecular Sets (MOSES): A Benchmarking Platform for Molecular Generation Models. *Front. Pharmacol.* **2020**, *11*, 565644. [CrossRef] [PubMed]
36. Kim, H.W.; Wang, M.; Leber, C.A.; Nothias, L.F.; Reher, R.; Kang, K.B.; van der Hooft, J.J.J.; Dorrestein, P.C.; Gerwick, W.H.; Cottrell, G.W. NPClassifier: A Deep Neural Network-Based Structural Classification Tool for Natural Products. *J. Nat. Prod.* **2021**, *84*, 2795–2807. [CrossRef] [PubMed]
37. Ertl, P.; Schuhmann, T. A Systematic Cheminformatics Analysis of Functional Groups Occurring in Natural Products. *J. Nat. Prod.* **2019**, *82*, 1258–1263. [CrossRef] [PubMed]
38. Zhang, J.; Mercado, R.; Engkvist, O.; Chen, H. Comparative Study of Deep Generative Models on Chemical Space Coverage. *J. Chem. Inf. Model.* **2021**, *61*, 2572–2581. [CrossRef] [PubMed]
39. Pahl, A.; Waldmann, H.; Kumar, K. Exploring Natural Product Fragments for Drug and Probe Discovery. *Chimia* **2017**, *71*, 653–660. [CrossRef]
40. Hanna, J.N.; Bekono, B.D.; Owono, L.C.; Toze, F.A.; Mbah, J.A.; Günther, S.; Ntie-Kang, F. A chemoinformatic analysis of atoms, scaffolds and functional groups in natural products. *Phys. Sci. Rev.* **2023**, *8*, 1341–1365. [CrossRef]

41. Vu, H.; Pedro, L.; Mak, T.; McCormick, B.; Rowley, J.; Liu, M.; Di Capua, A.; Williams-Noonan, B.; Pham, N.B.; Pouwer, R.; et al. Fragment-Based Screening of a Natural Product Library against 62 Potential Malaria Drug Targets Employing Native Mass Spectrometry. *ACS Infect. Dis.* **2018**, *4*, 431–444. [CrossRef]
42. Godinez, W.J.; Ma, E.J.; Chao, A.T.; Pei, L.; Skewes-Cox, P.; Canham, S.M.; Jenkins, J.L.; Young, J.M.; Martin, E.J.; Guiguemde, W.A. Design of potent antimalarials with generative chemistry. *Nat. Mach. Intell.* **2022**, *4*, 180–186. [CrossRef]
43. Bosc, N.; Felix, E.; Arcila, R.; Mendez, D.; Saunders, M.R.; Green, D.V.S.; Ochoada, J.; Shelat, A.A.; Martin, E.J.; Iyer, P.; et al. MAIP: A web service for predicting blood-stage malaria inhibitors. *J. Cheminform.* **2021**, *13*, 13. [CrossRef] [PubMed]
44. Probst, D.; Reymond, J.L. Visualization of very large high-dimensional data sets as minimum spanning trees. *J. Cheminform.* **2020**, *12*, 12. [CrossRef] [PubMed]
45. Murray, C.W.; Verdonk, M.L.; Rees, D.C. Experiences in fragment-based drug discovery. *Trends Pharmacol. Sci.* **2012**, *33*, 224–232. [CrossRef] [PubMed]
46. Woodhead, A.J.; Erlanson, D.A.; de Esch, I.J.; Holvey, R.S.; Jahnke, W.; Pathuri, P. Fragment-to-Lead Medicinal Chemistry Publications in 2022. *J. Med. Chem.* **2024**, *67*, 2287–2304. [CrossRef] [PubMed]
47. Chemical Computing Group. *Molecular Operating Environment (MOE)*; Chemical Computing Group: Montreal, QC, Canada, 2022.
48. Gargaro, A.R.; Soteriou, A.; Frenkiel, T.A.; Bauer, C.J.; Birdsall, B.; Polshakov, V.I.; Barsukov, I.L.; Roberts, G.C.; Feeney, J. The solution structure of the complex of Lactobacillus casei dihydrofolate reductase with methotrexate. *J. Mol. Biol.* **1998**, *277*, 119–134. [CrossRef] [PubMed]
49. Feeney, J.; Birdsall, B.; Kovalevskaya, N.V.; Smurnyy, Y.D.; Navarro Peran, E.M.; Polshakov, V.I. NMR structures of apo L. casei dihydrofolate reductase and its complexes with trimethoprim and NADPH: Contributions to positive cooperative binding from ligand-induced refolding, conformational changes, and interligand hydrophobic interactions. *Biochemistry* **2011**, *50*, 3609–3620. [CrossRef] [PubMed]
50. Grigalunas, M.; Burhop, A.; Christoforow, A.; Waldmann, H. Pseudo-natural products and natural product-inspired methods in chemical biology and drug discovery. *Curr. Opin. Chem. Biol.* **2020**, *56*, 111–118. [CrossRef] [PubMed]
51. Li, Y.; Cheng, S.; Tian, Y.; Zhang, Y.; Zhao, Y. Recent ring distortion reactions for diversifying complex natural products. *Nat. Prod. Rep.* **2022**, *39*, 1970–1992. [CrossRef]
52. Sorokina, M.; Merseburger, P.; Rajan, K.; Yirik, M.A.; Steinbeck, C. COCONUT online: Collection of Open Natural Products database. *J. Cheminform.* **2021**, *13*, 2. [CrossRef]
53. Zeng, T.; Liu, Z.; Zhuang, J.; Jiang, Y.; He, W.; Diao, H.; Lv, N.; Jian, Y.; Liang, D.; Qiu, Y.; et al. TeroKit: A Database-Driven Web Server for Terpenome Research. *J. Chem. Inf. Model.* **2020**, *60*, 2082–2090. [CrossRef]
54. Zeng, T.; Chen, Y.; Jian, Y.; Zhang, F.; Wu, R. Chemotaxonomic investigation of plant terpenoids with an established database (TeroMOL). *New Phytol.* **2022**, *235*, 662–673. [CrossRef] [PubMed]
55. Mendez, D.; Gaulton, A.; Bento, A.P.; Chambers, J.; De Veij, M.; Felix, E.; Magarinos, M.P.; Mosquera, J.F.; Mutowo, P.; Nowotka, M.; et al. ChEMBL: Towards direct deposition of bioassay data. *Nucleic Acids Res.* **2019**, *47*, D930–D940. [CrossRef] [PubMed]
56. Verras, A.; Waller, C.L.; Gedeck, P.; Green, D.V.; Kogej, T.; Raichurkar, A.; Panda, M.; Shelat, A.A.; Clark, J.; Guy, R.K.; et al. Shared Consensus Machine Learning Models for Predicting Blood Stage Malaria Inhibition. *J. Chem. Inf. Model.* **2017**, *57*, 445–453. [CrossRef] [PubMed]
57. Degen, J.; Wegscheid-Gerlach, C.; Zaliani, A.; Rarey, M. On the art of compiling and using drug-like chemical fragment spaces. *ChemMedChem* **2008**, *3*, 1503–1507. [CrossRef] [PubMed]
58. Bemis, G.W.; Murcko, M.A. The properties of known drugs. 1. Molecular frameworks. *J. Med. Chem.* **1996**, *39*, 2887–2893. [CrossRef] [PubMed]
59. Klein, G.; Kim, Y.; Deng, Y.; Senellart, J.; Rush, A.M. Opennmt: Open-source toolkit for neural machine translation. *arXiv* **2017**, arXiv:1701.02810.
60. Landrum, G. RDKit: A software suite for cheminformatics, computational chemistry, and predictive modeling. *Greg Landrum* **2013**, *8*, 5281.

molecules

Article

Water Exchange from the Buried Binding Sites of Cytochrome P450 Enzymes 1A2, 2D6, and 3A4 Correlates with Conformational Fluctuations

Olgun Guvench

Department of Pharmaceutical Sciences and Administration, School of Pharmacy, Westbrook College of Health Professions, University of New England, 716 Stevens Avenue, Portland, ME 04103, USA; oguvench@une.edu

Abstract: Human cytochrome P450 enzymes (CYPs) are critical for the metabolism of small-molecule pharmaceuticals (drugs). As such, the prediction of drug metabolism by and drug inhibition of CYP activity is an important component of the drug discovery and design process. Relative to the availability of a wide range of experimental atomic-resolution CYP structures, the development of structure-based CYP activity models has been limited. To better characterize the role of CYP conformational fluctuations in CYP activity, we perform multiple microsecond-scale all-atom explicit-solvent molecular dynamics (MD) simulations on three CYP isoforms, 1A2, 2D6, and 3A4, which together account for the majority of CYP-mediated drug metabolism. The MD simulations employ a variety of positional restraints, ranging from keeping all CYP atoms close to their experimentally determined coordinates to allowing full flexibility. We find that, with full flexibility, large fluctuations in the CYP binding sites correlate with efficient water exchange from these buried binding sites. This is especially true for 1A2, which, when restrained to its crystallographic conformation, is unable to exchange water between the binding site and bulk solvent. These findings imply that, in addition to crystal structures, a representative ensemble of conformational states ought to be included when developing structure-based CYP activity models.

Keywords: cytochrome P450; CYP; 1A2; 2D6; 3A4; binding site; conformation; flexibility; water; molecular dynamics

1. Introduction

Human cytochrome P450 enzymes (CYPs) are a major contributor to the metabolism of small-molecule pharmaceuticals (drugs) and metabolize ~75% of FDA-approved drugs [1,2]. CYPs accomplish this action by catalyzing oxidation reactions that convert hydrophobic substrates to hydrophilic products [3]. This conversion alters the pharmaceutical activity, leading to products that can be less or more active than the substrates [4], and generally favors the elimination of compounds from the body owing to either the direct enhancement of water solubility or the facilitation of subsequent conjugation reactions that further enhance water solubility [5]. In addition to drug metabolism by CYPs, there also exist the possibilities of drugs acting to inhibit or to induce CYP activity [6].

Important opportunities for the application of computational approaches in drug discovery and design are the prediction of the drug-specific impacts of and impacts on CYP activity. Experimental approaches typically entail the extraction of CYP-rich microsomes from hepatic tissue for assaying candidate drug molecules [7,8]. By reducing the need for such experimental assays, predictive computational models can optimize the time and materials costs associated with determining CYP activity in relation to a drug or drug candidate.

Two major categories of computational modeling for CYP activity/inhibition are "ligand-based" and "structure-based" [9]. Ligand-based models are developed using known structures and activities of drugs, which serve as a training set without consideration of the structure of the CYP enzymes themselves. In contrast, structure-based methods directly use the three-dimensional structures of CYP enzymes. "Machine learning", which has typically been used to support ligand-based modeling [9,10], has the capacity to simultaneously incorporate aspects associated with ligand-based and structure-based approaches within a single model [11], and therefore may be considered a third category. The success of ligand-based modeling is impressive, with the caveat that a limitation or bias is necessarily imposed by the training set. Such limitation/bias is relevant when attempting to evaluate compounds that do not have substantial chemical similarity to compounds included in the training, leading to missed predictions, or conversely, when the training set is very structurally diverse, leading to a general model that cannot be used to fine-tune structural changes during lead optimization [12]. Structure-based methods are conceptually attractive, as they do not suffer this training-set-associated limitation/bias and can also take into account drug chirality, unlike ligand-based models trained using 2D chemical structures [9]. However, the development of structure-based models has proven to be a challenge despite the availability of experimental atomic-resolution structures for CYPs [13]. Factors contributing to this challenge are the range of different human CYPs and their differing substrate binding sites; the buried nature of CYP binding sites; and the plasticity of the binding site of a given CYP, which can entail different binding site conformations that enable binding of structurally diverse substrates [14].

Toward addressing these three factors, we apply all-atom explicit-solvent molecular dynamics (MD) simulations to determine the binding site properties of three distinct CYPs, 1A2, 2D6, and 3A4, because these three CYP isoforms together metabolize ~50% of the CYP-metabolized drugs that are approved or in development [2]. Our analysis of the MD data focuses on the hydration properties of the CYP binding sites, with the aim of determining the extent of binding-site hydration in the absence of a ligand and the rate of exchange of water molecules between the binding site and bulk solvent, along with the dependence of these properties on the flexibility of the CYP protein. We find that restraining CYP atoms to their positions as solved using X-ray crystallography can slow down binding site water exchange with the bulk solvent, and very dramatically in the case of 1A2, whereas the full protein flexibility observed in non-restrained MD enables the efficient exchange of water molecules. Additionally, both the average volumes of the 1A2, 2D6, and 3A4 binding sites and the observed fluctuations in these volumes tend to be smaller in restrained simulations relative to non-restrained simulations, suggesting a link between enhanced structural fluctuations and water exchange. Our findings support the use of conformational ensembles determined using non-restrained MD simulations in developing structure-based models for predicting the binding of small molecules to CYPs.

2. Results and Discussion

2.1. CYP Binding Site Residues Are Flexible

The CYP 1A2, 2D6, and 3A4 crystal structure conformations all have buried binding sites characterized by a large pocket similarly located within the interiors of the proteins and defined by the heme moiety on one side. Fpocket [15] analysis (see Section 4) of these crystal structures identifies many pockets for each structure, although, upon visual inspection, for each protein crystal structure there is one obvious pocket corresponding to the binding site (Figure 1a–f). Per fpocket, the respective volumes for the 1A2, 2D6, and 3A4 pockets are 512 Å^3, 1019 Å^3, and 2206 Å^3, and these pockets are defined by the heme moiety plus 20, 34, or 46 amino acid residues (Table 1). In addition to the contrasting volumes and numbers of associated residues, the pockets for 2D6 and for 3A4 both contain portions that connect the binding sites to the exterior of the protein, whereas the pocket for 1A2 does not.

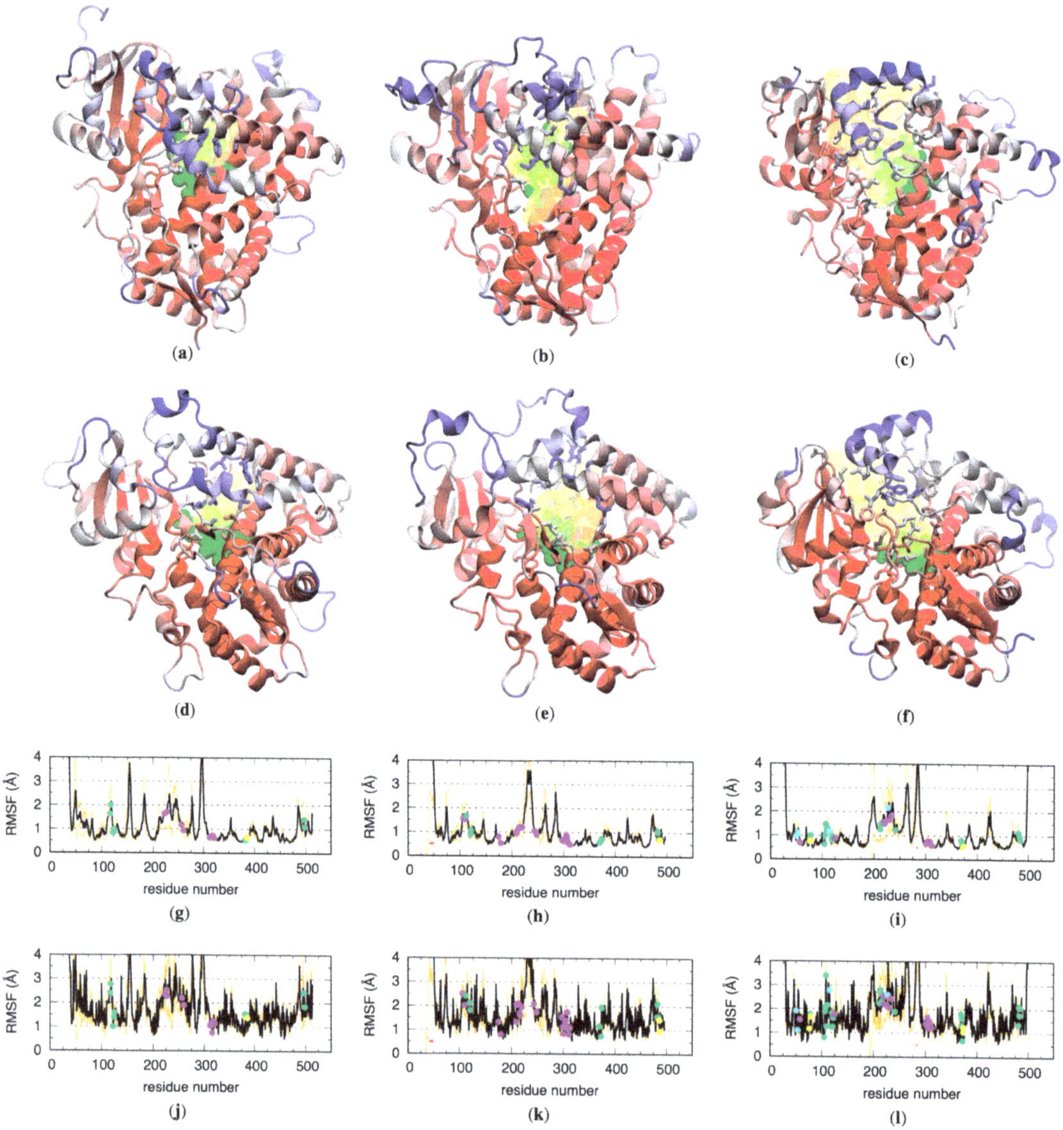

Figure 1. CYP crystallographic binding sites and their flexibility in non-restrained MD simulations. Molecular graphics of the 1A2, 2D6, and 3A4 crystal structures used as MD starting conformations are drawn from (**a–c**) the "distal face" and (**d–f**) the "side view" perspectives [16], with the binding site volume as determined using fpocket shown as a transparent yellow surface, the associated binding site residue (see Table 1) sidechains as sticks, and the heme group as green van der Waals spheres. The protein cartoon representation is colored according to data in the corresponding C^{α} root-mean-squared fluctuation (RMSF) graph (**g–i**), with increasing values from red through white and on to blue. Binding site residue sidechains are likewise colored according to data in the corresponding sidechain RMSF graph (**j–l**). Binding site residues are noted as points in (**g–l**) and are colored according to the STRIDE [17] secondary structure computed from the corresponding crystal structure conformation (AlphaHelix = purple, 310Helix = blue, Strand = yellow, Coil/Turn/Bridge = green), and residues missing in the crystal structures are noted with a red bar at $y = 0.5$ Å. RMSFs are averaged across all

non-restrained MD trajectories for a given CYP (see Section 4) and are drawn as black lines, with orange lines at +/− one standard deviation. For clarity, the range of color red through blue for the molecular graphics is determined using the minimum overall RMSF for the minimum value and the maximum binding site residue RMSF for the maximum value from the corresponding graph.

Table 1. Crystallographic binding site volumes and residues as determined with fpocket.

CYP	PDB ID	Binding Site Pocket Volume (Å^3)	Binding Site Residue Numbers [1]
1A2	2HI4	512	117 118 122 124 125 223 226 227 256 260 312 313 316 317 320 321 382 386 497 498 900
2D6	2F9Q	1019	106 110 112 120 121 175 179 209 210 213 214 216 217 220 244 248 297 300 301 304 305 307 308 309 311 312 370 373 374 482 483 484 486 487 600
3A4	1TQN	2206	50 53 57 76 78 79 105 106 107 108 109 111 115 119 120 121 122 125 212 213 215 216 220 221 223 224 227 230 234 241 301 304 305 308 309 312 369 370 371 372 373 374 481 482 483 484 508

[1] Last residue number in each list is the heme moiety.

All-atom explicit-solvent MD simulations started from the crystallographic conformations showed that binding site residues include atoms with high flexibility. To minimize bias introduced by equilibration of the solvent causing perturbation of the crystallographic conformations, each MD simulation entailed five successive 250 ns stages that created a single, continuous 1250 ns trajectory: Stage 1 had strong positional restraints on all non-hydrogen protein atoms, Stage 2 had weak positional restraints on all non-hydrogen protein atoms, Stage 3 had weak positional restraints on C^{α} protein atoms only, and Stage 4.1 and Stage 4.2 had no positional restraints on any individual atoms (see Section 4). As expected, high root-mean-squared fluctuation (RMSF) values of the protein backbone C^{α} and sidechain atoms from the non-restrained (Stage 4.1 and 4.2) portions of these trajectories were associated with loop and N-terminal regions of these proteins due to their lack of α-helical or β-strand secondary structures (Figure 1a–f). Interestingly, high RMSF values were also associated with those atoms in α-helical portions of the proteins that contained binding site residues (Figure 1a–l), in contrast with other α-helical portions of the proteins, which tended to have the lowest RMSF values. This high degree of flexibility in binding site residues with crystallographic secondary structures is presumably an enabling factor in allowing either substrate access to or product release from the 1A2, 2D6, and 3A4 binding sites, or in allowing the binding site for a given CYP to bind a diverse array of substrates.

2.2. *Conformational Flexibility Is Required for Binding Site Water Access in CYP 1A2 but Not in CYP 2D6 or CYP 3A4*

The restrained stages of the MD trajectories highlight the variable importance of protein flexibility in enabling access to the buried binding sites of CYP 1A2, 2D6, and 3A4. Because the current MD data are for systems without small molecule substrates, surveying binding site water molecule hydrogen-bond networks can provide a proxy for binding site accessibility from the bulk solvent, with the caveat that water molecules and drug molecules may use different routes to transit between the bulk solvent and a binding site [18–20].

In the case of CYP 1A2, the creation of a bulk solvent-connected hydrogen-bond network of water molecules is not possible when the protein is restrained to its crystallographic conformation during the Stage 1 MD. In Stage 1, an isolated cluster of water molecules exists adjacent to the heme iron. It is only with the introduction of full sidechain flexibility in Stage 3 of the CYP 1A2 simulations that a substantial ratio, ~30% (Figure 2a,d), of the MD snapshots evince a connected network composed of contiguous hydrogen-bonded water molecules spanning the binding site to the bulk solvent. Furthermore, it is only when CYP 1A2 is non-restrained in Stage 4.1 and Stage 4.2, and therefore allowed to sample its natural ensemble of conformational states, that the large majority of individual conformations (MD snapshots) contain this connected network (Figure 2a,d).

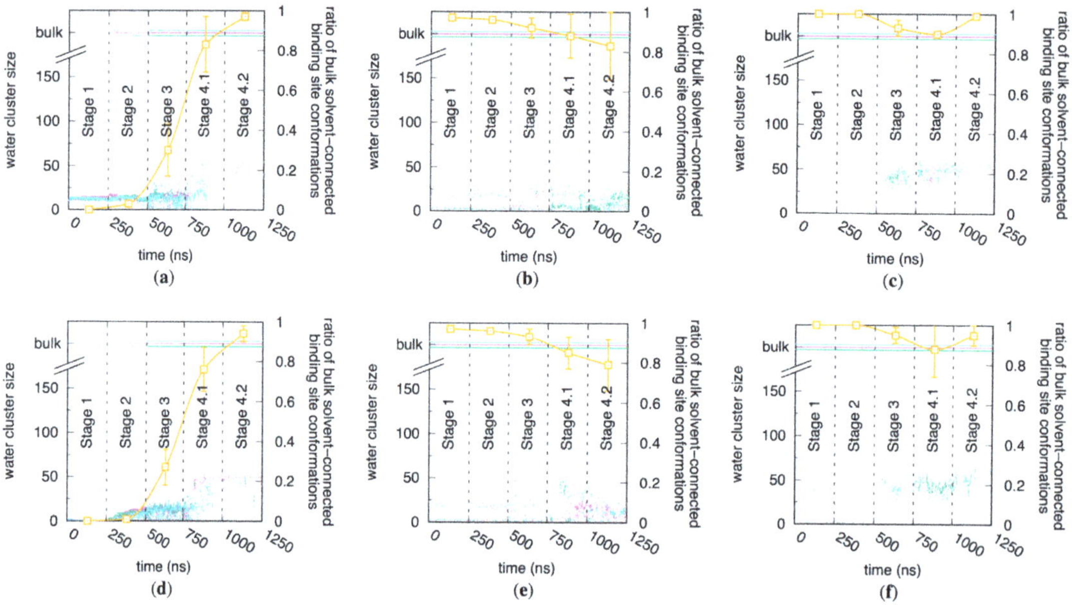

Figure 2. Connection of CYP binding sites to the bulk solvent as a function of increasing protein flexibility. Data are for 1A2, 2D6, and 3A4 using (**a–c**) the "2 Å" and (**d–f**) the "3 Å" solvation protocols (see Section 4). Purple/green/blue dots are the water cluster size time series data from triplicate MD simulations for each CYP/protocol combination, with water cluster size computed as described in Section 4; data for clusters connected to the bulk solvent are plotted at values of "bulk," "bulk−", and "bulk+" for clarity. The per-stage ratios of bulk solvent-connected binding site conformations, plotted in orange, are derived from these data, with ratios computed for each stage of each run and then averaged across the three runs; error bars are the standard deviation in the three-run average ratio for each stage. In a time series, a binding site conformation is considered to be bulk solvent-connected if its corresponding water cluster size = "bulk" (see Section 4).

This same analysis for CYP 2D6 and CYP 3A4 reveals that, despite being homologs of 1A2, these two proteins in their crystal structure conformations can accommodate a connected network of water molecules between the binding site and bulk solvent (Figure 2b,e and c,f). The large majority of individual conformations across all of the MD stages for 2D6 and 3A4 have such a hydrogen-bonded water network, including Stage 1, in which CYP atom coordinates are restrained to their crystallographic values. In fact, this is the case for nearly all sampled conformations in Stage 1 and Stage 2 of the CYP 2D6 and 3A4 simulations, which is opposite the behavior of CYP 1A2. Interestingly, with the introduction of full protein flexibility in Stage 4.1, these two proteins sample conformations with binding sites that are cut off from the bulk solvent. It is tempting to speculate that these conformational fluctuations occurring in non-restrained MD may enhance catalysis by enabling full isolation of the binding site from the bulk solvent and stronger binding of substrates in poses with maximized protein–ligand interactions.

2.3. Binding Site Volumes Increase with Increased Protein Flexibility and Encompass Increasing Numbers of Water Molecules

Increased CYP flexibility enables increased binding site volumes per fpocket analysis of the individual snapshots across all MD stages. To perform the analysis, the residues in

Table 1 were used as the pocket definition input for fpocket. Distributions of the fpocket-computed volumes based on these pocket definitions and aggregated on a per-replicate and per-stage basis generally show a shift in distribution peaks to larger values in the progression from Stage 1 to Stage 4.x (Figure 3). The most dramatic change occurs in the 1A2 simulations, wherein the shift between the most probable volume sampled when the protein is restrained to its crystallographic conformation in Stage 1 to that sampled during the last, non-restrained stage of MD in Stage 4.2 is the difference between 500 $Å^3$ and 1400 $Å^3$—a 900 $Å^3$, or 180%, increase (Figure 3a). The largest such increases for 2D6 and for 3A4 are 50% (1200 $Å^3$ to 1800 $Å^3$, Figure 3b) and 15% (2600 $Å^3$ to 3000 $Å^3$, Figure 3f), respectively. In addition to the increase in binding site pocket volumes, the non-restrained simulations exhibit increased fluctuations in binding site pocket volumes relative to the restrained simulations. This is most apparent when looking at the peak heights of the normalized probability distributions for these volumes: distributions from the Stage 4.x MD have significantly lower peaks and wider bases than those from the Stage 1 and 2 MDs for all three CYPs (Figure 3). Prior MD simulation of CYP 2D6 noted that the volume of its active site can fluctuate more than 50%, and this is true of the data from the present MD (Figure 3b,e) [21].

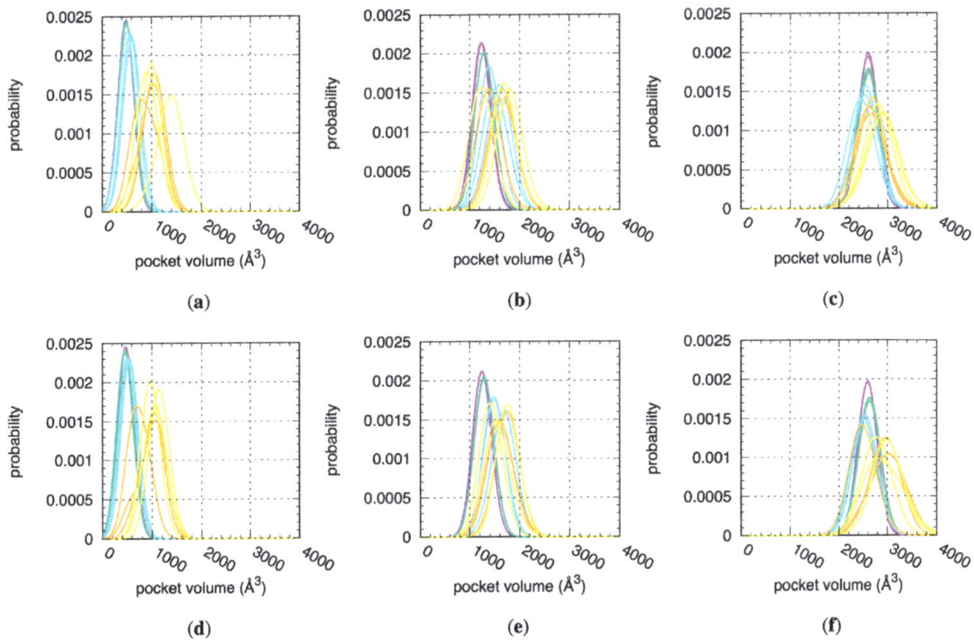

Figure 3. Distributions of fpocket-computed binding site volumes across all MD simulations. Data are for 1A2, 2D6, and 3A4 using (**a–c**) the "2 Å" and (**d–f**) the "3 Å" solvation protocols (see Section 4). Data for each distribution are from a single replicate for a single stage. Purple = Stage 1, green = Stage 2, blue = Stage 3, orange = Stage 4.1, yellow = Stage 4.2.

Not surprisingly, similar trends exist for the number of water molecules within the binding site pockets (Figure 4). For this analysis, for a given MD snapshot, a water molecule was taken to be within the binding site pocket volume if the oxygen atom of that water molecule was located within the sphere radius of any of the pocket alpha spheres computed using fpocket for that snapshot. The Stage 4.x MD data therefore imply that, in solution at physiological temperature, CYP 1A2, 2D6, and 3A4 in their substrate- or inhibitor-free forms sample well-hydrated binding site conformations that are enlarged relative to the crystallographic conformations. We note that a previous MD study found the 3A4 binding site to be occupied by 54–58 water molecules, whereas the present work puts this number at 110–120 (Figure 4c,f), and this may be explained by the 4 ns timescale of the production MD in that work versus the microsecond timescale MD here as well as by differences in how the binding site pocket was defined [22].

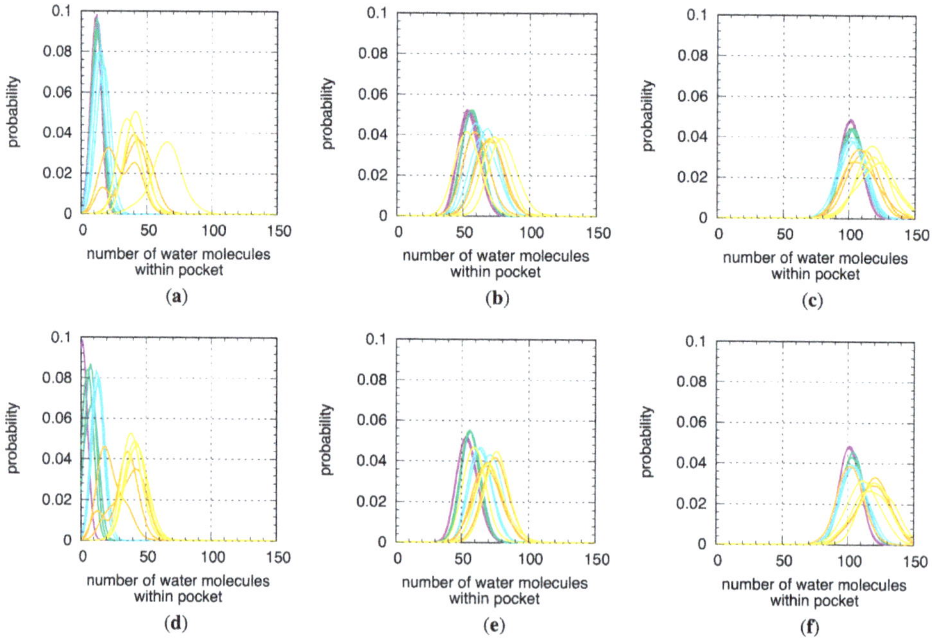

Figure 4. Distributions of the number of water molecules within fpocket-computed binding site volumes across all MD simulations. Data are for 1A2, 2D6, and 3A4 using (**a**–**c**) the "2 Å" and (**d**–**f**) the "3 Å" solvation protocols (see Section 4). Data for each distribution are from a single replicate for a single stage. Purple = Stage 1, green = Stage 2, blue = Stage 3, orange = Stage 4.1, yellow = Stage 4.2.

2.4. Protein Flexibility Is Especially Important for Binding Site Water Exchange in CYP 1A2

The equilibrium data from the MD simulations show that for CYP 1A2, protein flexibility is required to connect water molecules in the binding site with those in the bulk solvent via contiguous hydrogen-bond networks, but it is not required for CYP 2D6 and 3A4 (Figure 2). However, these data do not speak to the ability of water molecules within CYP binding sites to exchange with the bulk solvent external to the protein. That is to say, a particular water molecule within the binding site may participate in a hydrogen-bond network of water molecules connecting with the bulk solvent, yet this water molecule may still be effectively "stuck" within the binding site in terms of the timescales required for it to diffuse out into the bulk solvent. Correlation analysis of the residence of CYP binding site water molecules provides a kinetic perspective and clearly demonstrates that CYP flexibility correlates with the ability of water molecules in the binding site to move out into the bulk solvent and vice versa.

Correlation analysis was performed on a per-replicate, per-stage basis by comparing the identities of the water molecules within the binding site at time t and time $t + \Delta t$. Each water molecule in the system was assigned a unique identifier, which it retained for the duration of the MD simulation. The correlation $C(\Delta t)$ is defined as:

$$C(\Delta t) := \left\langle \frac{n_{\text{persistent}}(t, t+\Delta t)}{n_{\text{persistent}}(t, t+\Delta t) + n_{\text{transient}}(t, t+\Delta t)} \right\rangle_{t=0\text{ ns}\ldots(250\text{ ns}-\Delta t)}.$$

Here, $n_{\text{persistent}}(t, t + \Delta t)$ is the count of the water molecules within the binding site pocket volume at time t that persist in the binding pocket volume at time $t + \Delta t$ based on their having the same unique identifiers. $n_{\text{transient}}(t, t + \Delta t)$ is the count of the water molecules within the binding site pocket volume at time t that are not in the binding site pocket volume at time $t + \Delta t$ summed with the count of the water molecules that are in the binding site pocket volume at time $t + \Delta t$ but not in the binding site pocket volume at time t. The angle brackets indicate an average taken over all t in the interval 0 ns to (250 ns—Δt). Identical to the binding site water analysis in the previous subsection, a water molecule was taken to be within the binding site pocket volume if the oxygen atom of that water molecule was located within the sphere radius of any of the pocket alpha spheres computed with fpocket for that snapshot. Based on this definition for the correlation, $C(\Delta t = 0 \text{ ns}) = 1$ because $n_{\text{transient}} = 0$ and $C(\Delta t) = 0$ when there is a complete exchange of water molecules in the binding pocket between time t and $t + \Delta t$, that is, when $n_{\text{persistent}} = 0$.

Results of this kinetic analysis are similar to the above equilibrium analysis. In the case of CYP 1A2, there is a dramatic dependence of the kinetics of water exchange on protein flexibility, with greater flexibility promoting more rapid exchange (Figure 5a,d), parallel to greater flexibility promoting equilibrium hydrogen-bond networks between the binding site water and bulk solvent (Figure 2a,d). While similar trends exist for the relationship between increased protein flexibility and increased water exchange kinetics for CYP 2D6 (Figure 5b,e) and CYP 3A4 (Figure 5c,f), they are less dramatic, and there are instances of replicate simulations wherein faster exchange occurs in the restrained (i.e., less flexible) simulation stages (Stages 1–3; purple, green, and blue traces in Figure 5b,c,e,f) compared with the non-restrained simulation stages (Stages 4.1–4.2; orange and yellow traces in the same Figure 5 panels). In strong contrast, for CYP 1A2, the kinetics of water exchange are always in the same order: fully restrained protein with strong positional restraints (Stage 1) < fully restrained protein with weak positional restraints (Stage 2) < C^{α} atom-only weak positional restraints (Stage 3) < non-restrained protein (Stages 4.1 and 4.2). Taken together, this suggests that the capacity to form a contiguous network of hydrogen-bonded water molecules connecting the binding site and the bulk solvent is a necessary condition for efficient water exchange.

The jitter in the CYP 1A2 Stage 1 traces in Figure 5d arises from the combination of the lack of water exchange with the bulk solvent and the "3 Å" protocol, which results in $n_{\text{persistent}} = 1$ and $n_{\text{transient}} = 0$ for over 80% of the snapshots across the three replicates. This lone water molecule is the one that is initially ligated to the heme iron and that retains this ligation status for the duration. Per the force field model, this ligation is a non-covalent interaction. That is, there is no bonded term between the water molecule and the iron atom; rather, the association is maintained solely through electrostatic and Lennard-Jones interaction terms. Specifically, the strong electrostatic interaction between the large positive charge of +1.24 at the iron[III] site (see Supplementary Materials) and the large negative charge of −0.834 at the TIP3P water oxygen site [23] ensures persistence of this water molecule as the sixth ligand to the heme iron.

Potentially relevant to the differences observed between 1A2 versus 2D6 and 3A4 may be the crystal structures used as starting conformations for the simulations. As all simulations in this study were of these proteins in their substrate-/inhibitor-free forms, corresponding crystal structures were used if available. While this was possible for 2D6 and 3A4, it was not for 1A2 (see Section 4). Thus, there exists the possibility that the 1A2 results reflect the 1A2 crystal conformation being optimal for binding of the alpha-naphthoflavone ligand present in the 2HI4 co-crystal. As such, removing alpha-naphthoflavone and using the remaining 1A2 protein coordinates to start the simulations may have yielded different results than using 1A2 coordinates from a substrate-/inhibitor-free crystal structure, were such a structure available. In the context of the bound substrate and oxygen, residual water in the active site may lead to uncoupling of the chemistry, which in turn would produce water rather than the desired oxidized product [24,25]. Therefore, the lack of water connection and exchange between the 1A2 binding site and bulk solvent for the conformationally restrained portions (Stages 1–3) of the 1A2 MD trajectories may be a reflection of a protein conformation optimized to take this factor into account. A definitive answer waits on the availability of a substrate-/inhibitor-free crystal structure for human 1A2. We refer interested readers to the discussions of the "solvent channel" and the "water channel" across various CYPs in [18]; access/egress channels in CYPs and controlled exposure of water to the binding site in [19]; and evacuation of water molecules from the 2D6 binding site to accommodate ligand binding in [26].

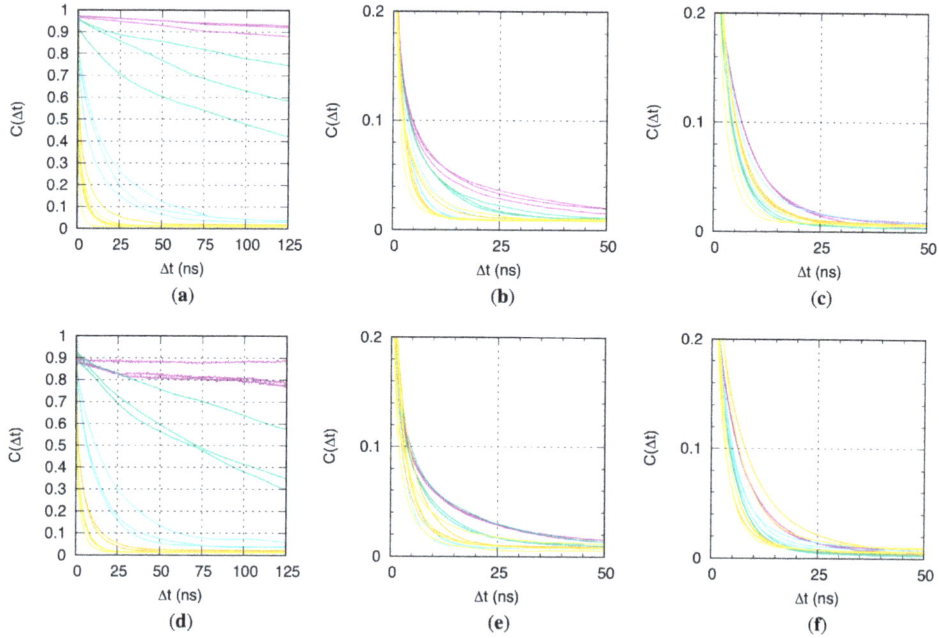

Figure 5. CYP binding site water molecule residence correlations $C(\Delta t)$. Data are for 1A2, 2D6, and 3A4 using (**a**–**c**) the "2 Å" and (**d**–**f**) the "3 Å" solvation protocols. Purple = Stage 1, green = Stage 2, blue = Stage 3, orange = Stage 4.1, yellow = Stage 4.2.

3. Conclusions

If structure-based models of CYP activity can be created to accurately predict drug binding, including both the binding pose and the binding strength, then it will be possible to confidently determine not only if a compound will be bound, but also if it will be oxidized based on the location of the reactive functional groups in the bound pose relative to the CYP heme moiety. For compounds that are predicted to bind but whose binding poses are incompatible with oxidation by the CYP heme, the predicted strength of binding may be correlated with these compounds' potential to act as CYP inhibitors, which is also an important consideration in drug development. Additionally, knowledge of the inhibitor binding poses can help inform synthetic modifications to convert a drug candidate that is a CYP inhibitor to a new compound that is not as part of the computational drug discovery and design process. Critical to the construction of successful structure-based models is an accurate and complete accounting of the thermodynamically important conformations for a given CYP, because the existence of such a conformational ensemble enables a given CYP to bind a variety of structurally different small molecules [27].

Non-restrained microsecond-scale non-biased all-atom explicit-solvent MD simulations, like the ones in the present study, have the potential to enable the enumeration of thermodynamically important protein conformations. However, as with any methodology, possible limitations must be kept in mind. Chief among these are the accuracy of the force field and the completeness of the sampling. With respect to the force field, one factor is the variable performance of different protein force fields, for example, as pertains to modeling well-structured versus flexible or disordered proteins [28–31]. Another is the choice of water force field parameters, which must be in balance with the protein force field parameters to correctly capture protein–water and water–water interactions but which may not accurately capture the structural and transport properties of the water itself [23,32,33]. A third is the parametrization of the iron[III]-containing heme moiety, which in the present work had parameters assigned by analogy; it may be worthwhile to use quantum mechanics calculations to validate or optimize such parameters. A fourth factor is the lack of dynamic polarizability or charge transfer inherent in fixed-charge force fields, though the use of polarizable force fields does not guarantee that the first three factors will not be issues and comes with additional computational expense [34]. With respect to the completeness of sampling, the timescale of the simulations may pose a limitation. Related to this is the choice of starting conformation, including the necessity to computationally construct missing residues, because the conformations sampled during a simulation may correlate more or less strongly with the starting conformation depending on the timescale of the simulation. Despite these possible limitations, simulations like the ones here offer an opportunity to develop atomic-resolution conformational ensembles beyond what is possible using experimental approaches.

Conformational ensembles of CYP 2D6 from MD simulations have recently been used successfully in the context of machine-learning models of CYP inhibition [11]. In that work, conformations were extracted from 3 ns length trajectories that employed an implicit solvent model. It remains to be seen whether model accuracy could be improved beyond the reported accuracy of 75% by incorporating conformations extracted from microsecond-scale explicit solvent simulations like those in the present work, especially as there exist experimental data to support the idea that human CYPs bind drugs and other substrates predominantly through conformational selection [27]. An open question is whether the simulations used to generate such conformational ensembles ought to incorporate lipid bilayers. CYP crystal structures are bilayer-free because they employ truncated constructs that exclude the transmembrane domain and retain only the globular catalytic domain, whereas the membrane bilayer may affect CYP structure, drug binding, and interactions with redox partners [35]. Recent pioneering work on CYP 2D6 has demonstrated stable microsecond-scale MD simulation of its membrane-anchored form, which was created by combining crystallographic coordinates of the catalytic domain with modeled coordinates for the transmembrane domain embedded in a bilayer [36]. Extension of this work led to the observation of spontaneous ligand binding to CYP 2D6, including facilitated ligand update by an allelic variant known to have increased metabolic activity [26]. It is an open question whether conformational fluctuations in the globular catalytic domain of CYP 2D6 are strongly affected by the presence of the transmembrane domain and its embedding within a bilayer; if they are, it will be important to sample conformations from simulations with lipid bilayers to best incorporate the conformational diversity of a particular CYP during development of a structure-based model. Additionally, the findings regarding the influence of the bilayer on the conformational ensemble of the globular catalytic domain for a given CYP will likely not be generalizable, as it appears that even two CYP proteins in the same subfamily—2C9 and 2C19—can have differing bilayer interactions [37].

4. Methods

4.1. Molecular Dynamics (MD) Simulations

X-ray crystal structures of the human cytochrome P450 enzymes 1A2 [38], 2D6 [39], and 3A4 [16] were used for starting coordinates for the MD simulations. The PDB [40] IDs for these structures are 2HI4, 2F9Q, and 1TQN, respectively. 2HI4 was chosen because it is the only structure of human 1A2 in the PDB. 2F9Q was chosen because it is the only ligand-free structure of human 2D6 in the PDB. 1TQN was chosen because it is a ligand-free structure of human 3A4, was solved at a better resolution and with fewer missing residues than the previously solved ligand-free structure (PDB ID 1W0E [41]), and was solved at a better resolution than a more recently solved ligand-free structure (PDB ID 4I3Q [42]). Chain A coordinates from each crystal structure were processed with the Reduce software [43] v. 3.23.130521 to add explicit hydrogen atoms and to optimize Asn and Gln sidechain amide group orientations; OH, SH, NH_3^+, and methionine methyl rotations; and histidine sidechain protonation states. ROSETTA3 [44] v. 2023.26.351 was used to model the missing loop residues, which were 42–51 for 2F9Q and 282–285 for 1TQN, via the RosettaRemodel functionality [45], and residues immediately flanking the missing residues were also included in the loop modeling. Missing N- and C-terminal residues, that is, those not located in the X-ray experiment, were excluded. Using the CHARMM software [46] v. c45b1, N- and C-termini were constructed in their ionized forms appropriate for neutral pH, and the finalized protein and heme starting coordinates were solvated in cubes of explicit water molecules containing 140 mM NaCl and sufficient additional Cl^- ions to generate net neutral systems. First, the cube of water molecules was added, followed by deletion of water molecules that overlapped CYP atoms, and finally, Na^+ and Cl^- ions were added through the random replacement of water molecules located at least 6 Å away from the protein. Prior to solvation, CYP atoms were oriented such that the heme iron was at the origin, the plane of the heme molecule was in the xy plane, and the Cys residue ligated to the heme iron below the xy plane. The cubic systems had edge lengths of l + 20 Å, where l is length of the longest x-, y-, or z-span of the protein after orientation.

Owing to the buried catalytic pockets of the CYP proteins, two separate solvation protocols were applied to each protein structure to test the effect of the number of water molecules initially within the catalytic pockets on the simulation results. In the first protocol, "2 Å," water molecules whose oxygen atoms were within 2 Å of any CYP atoms were deleted. In the second protocol, "3 Å," water molecules whose oxygen atoms were within 3 Å of any CYP atoms were deleted.

MD simulations for each protein were performed in triplicate. For 1A2 (2HI4), starting coordinates for all three simulations were identical. For 2D6 (2F9Q) and 3A4 (1TQN), starting coordinates were the three best-scored RosettaRemodel outputs each. The triplicate simulations combined with the two different solvation protocols meant six independent MD simulation trajectories per CYP. OpenMM [47] v. 8.0.0 was used for all MD simulations, including initial energy minimization and heating. Periodic boundary conditions with a 9 Å cutoff were applied [48], and particle-mesh Ewald was used to account for electrostatic [49] and Lennard-Jones contributions [50] beyond the cutoff. Heating was performed across 0.1 ns from an initial temperature of 31 K, used for assignment of initial randomized velocities, to the target temperature of 310 K using Langevin integration [51] with a friction coefficient of 1 ps^{-1}, constant system volume, and harmonic positional restraints of the form $k \times (\Delta r)^2$ on all CYP non-hydrogen atoms, where Δr is the displacement from the starting coordinates for a given atom and k = 1 $kcal \cdot mol^{-1} \cdot Å^{-2}$. This was followed by five successive stages of production MD simulations, all run at 310 K using Langevin integration with a friction coefficient of 0.1 ps^{-1} and at a constant system pressure of 1.01325 bar using a Monte Carlo barostat [52,53]. For Stage 1, the harmonic positional restraints were the same as those used for heating. For Stage 2, k was reduced to 0.1 $kcal \cdot mol^{-1} \cdot Å^{-2}$. For Stage 3, k was maintained at 0.1 $kcal \cdot mol^{-1} \cdot Å^{-2}$, but positional restraints were removed from all atoms except for CYP protein C^{α} atoms. For Stages 4.1 and 4.2, all atomic positional restraints were removed, and a single positional restraint of the same form was applied to the center of mass of the CYP atoms, with k = 0.1 $kcal \cdot mol^{-1} \cdot Å^{-2}$. The purpose of the center-of-mass restraint was simply to facilitate post-run analysis of the MD snapshots by preventing diffusion of the protein away from the center of the periodic system; without the restraint, each snapshot would need to be re-imaged in order to analyze the CYP–water interactions. Bonds involving hydrogen atoms and water geometries were constrained to their equilibrium values [54–56], allowing for the use of a 2 fs integration timestep. Each stage was 250 ns (125×10^6 timesteps) in length, and each successive stage was a continuation of the prior stage, accomplished by using the final atomic positions and velocities from the prior stage, thereby producing a continuous 1250 ns MD trajectory with gradually decreasing atomic positional restraints on the CYP atoms.

Snapshots were taken every 250 ps, resulting in 1000 snapshots per stage. Post-run analysis of snapshots was performed using the CHARMM program, the VMD [57] v. 1.9.4a57-arm64-Rev12 program, the fpocket [15] v. 4.1 program, as well as custom Bash and Python scripts. Crystallographic and MD snapshot conformations used as inputs for fpocket included hydrogen atoms and heme moieties, and fpocket was run with default parameter settings.

4.2. Force Field

The CHARMM 36m additive force field for proteins [28–31] v. jul20 (available at https://mackerell.umaryland.edu/charmm_ff.shtml, accessed on 15 January 2024) was used for all simulations. Water molecules were represented using the TIP3P model [23] modified for the CHARMM force field [58], and Na^+ and Cl^- ion parameters were as described in [59,60]. Force field parameters for the heme with Cys sidechain liganded in the thiolate form were directly transferred from existing parameters, with the following exceptions: the heme iron oxidation state was modeled as iron[III] for consistency with experimental data on CYPs having water as the sixth heme ligand [41], and this required changes to the heme partial charges as well as to the Lennard-Jones parameters for the heme iron atom [61]. Please refer to the Supplementary Materials for the complete set of parameters used in addition to v. jul20.

4.3. Root-Mean-Squared Fluctuation (RMSF) Analysis

Atomic RMSF values were first computed on a per-stage basis for each Stage 4.1 and Stage 4.2 portion of each trajectory. All 1000 snapshots from a stage were root-mean-square (RMS) aligned to the starting (crystallographic) conformation using C^{α} atom positions, and RMSF values for all C^{α} atoms were computed for this 1000-conformation ensemble. The same procedure was performed a second time, but using for the RMS alignment only those C^{α} atom positions with a previously computed RMSF value less than 2 Å, so that structural alignment to the reference conformation was not skewed by highly mobile atoms. RMSF values for all C^{α} atoms and for all sidechain atoms were computed based on this second alignment. A single RMSF value per sidechain was computed by summing the squares of the RMSF values for all atoms in the sidechain, dividing this sum by the number of atoms in the sidechain, and then taking the square root of this sum.

RMSF values were reported in aggregate as residue-wise averages and standard deviations for each of the three CYP proteins. Data aggregation entailed averaging across twelve sets of RMSF data, each from a 1000 snapshot ensemble: ("2 Å" + "3 Å" solvation protocols) × (Stage 4.1 + Stage 4.2) × (triplicate MD simulations).

4.4. Binding Site Water Cluster Size

The water cluster size was defined as the number of water molecules within a hydrogen-bond network that included the water molecule ligated to the heme iron atom in the CYP binding site. A "hydrogen bond" was defined as two water molecules having their oxygen atoms within 3.5 Å of each other; that is, each water molecule in the cluster was hydrogen bonded to at least one other water molecule in the cluster. The iron-ligated water molecule was defined as the one having its oxygen atom within 3.5 Å of the heme iron. This water molecule was found to exist in all 30,000 MD snapshots across the six CYP/protocol combinations, with the exception of 4 snapshots: the Stage 1 first snapshot in two runs and the Stage 1 first two snapshots in the third run of the 2D6 simulations using the "3 Å" protocol. Cluster sizes as large as 79 were observed across the 18 MD trajectories, and these corresponded to binding site conformations not connected to the bulk solvent, while the next-smallest observed water cluster size beyond 79 water molecules contained more than

15,000 water molecules (water cluster size = "bulk"), indicative of bulk solvent-connected binding site conformations.

Supplementary Materials: The following supporting information can be downloaded at: https://www.mdpi.com/article/10.3390/molecules29020494/s1, patch file "toppar_all36_prot_heme.str.patch" containing CYP heme-specific topology and parameter information. Apply "toppar_all36_prot_heme.str.patch" to the default "toppar_c36_jul20/stream/prot/toppar_all36_prot_heme.str" file from the CHARMM 36 additive force field v. jul20 available at https://mackerell.umaryland.edu/charmm_ff.shtml, accessed on 15 January, 2024, e.g.: "patch toppar_c36_jul20/stream/prot/toppar_all36_prot_heme.str < toppar_all36_prot_heme.str.patch".

Funding: This research received no external funding.

Institutional Review Board Statement: Not applicable.

Informed Consent Statement: Not applicable.

Data Availability Statement: Data are contained within the article and Supplementary Materials.

Conflicts of Interest: The author declares no conflict of interest.

References

1. Esteves, F.; Rueff, J.; Kranendonk, M. The Central Role of Cytochrome P450 in Xenobiotic Metabolism—A Brief Review on a Fascinating Enzyme Family. *J. Xenobiotics* **2021**, *11*, 94–114. [CrossRef]
2. Rendic, S.; Guengerich, F.P. Survey of Human Oxidoreductases and Cytochrome P450 Enzymes Involved in the Metabolism of Xenobiotic and Natural Chemicals. *Chem. Res. Toxicol.* **2015**, *28*, 38–42. [CrossRef]
3. Guengerich, F.P. Cytochrome P450 and Chemical Toxicology. *Chem. Res. Toxicol.* **2008**, *21*, 70–83. [CrossRef]
4. Zanger, U.M.; Schwab, M. Cytochrome P450 enzymes in drug metabolism: Regulation of gene expression, enzyme activities, and impact of genetic variation. *Pharmacol. Ther.* **2013**, *138*, 103–141. [CrossRef]
5. Zhao, M.; Ma, J.; Li, M.; Zhang, Y.; Jiang, B.; Zhao, X.; Huai, C.; Shen, L.; Zhang, N.; He, L.; et al. Cytochrome P450 Enzymes and Drug Metabolism in Humans. *Int. J. Mol. Sci.* **2021**, *22*, 12808. [CrossRef]
6. Hakkola, J.; Hukkanen, J.; Turpeinen, M.; Pelkonen, O. Inhibition and induction of CYP enzymes in humans: An update. *Arch. Toxicol.* **2020**, *94*, 3671–3722. [CrossRef]
7. Fowler, S.; Zhang, H. In vitro evaluation of reversible and irreversible cytochrome P450 inhibition: Current status on methodologies and their utility for predicting drug-drug interactions. *AAPS J.* **2008**, *10*, 410–424. [CrossRef]
8. Rao Gajula, S.N.; Pillai, M.S.; Samanthula, G.; Sonti, R. Cytochrome P450 enzymes: A review on drug metabolizing enzyme inhibition studies in drug discovery and development. *Bioanalysis* **2021**, *13*, 1355–1378. [CrossRef]
9. Olsen, L.; Oostenbrink, C.; Jørgensen, F.S. Prediction of cytochrome P450 mediated metabolism. *Adv. Drug Deliv. Rev.* **2015**, *86*, 61–71. [CrossRef]
10. Zhai, J.; Man, V.H.; Ji, B.; Cai, L.; Wang, J. Comparison and summary of in silico prediction tools for CYP450-mediated drug metabolism. *Drug Discov. Today* **2023**, *28*, 103728. [CrossRef]
11. Martiny, V.Y.; Carbonell, P.; Chevillard, F.; Moroy, G.; Nicot, A.B.; Vayer, P.; Villoutreix, B.O.; Miteva, M.A. Integrated structure- and ligand-based in silico approach to predict inhibition of cytochrome P450 2D6. *Bioinformatics* **2015**, *31*, 3930–3937. [CrossRef]
12. Kato, H. Computational prediction of cytochrome P450 inhibition and induction. *Drug Metab. Pharmacokinet.* **2020**, *35*, 30–44. [CrossRef]
13. Dong, D.; Wu, B.; Chow, D.; Hu, M. Substrate selectivity of drug-metabolizing cytochrome P450s predicted from crystal structures and in silico modeling. *Drug Metab. Rev.* **2012**, *44*, 192–208. [CrossRef]
14. Nair, P.C.; McKinnon, R.A.; Miners, J.O. Cytochrome P450 structure–function: Insights from molecular dynamics simulations. *Drug Metab. Rev.* **2016**, *48*, 434–452. [CrossRef]
15. Le Guilloux, V.; Schmidtke, P.; Tuffery, P. Fpocket: An open source platform for ligand pocket detection. *BMC Bioinform.* **2009**, *10*, 168. [CrossRef]
16. Yano, J.K.; Wester, M.R.; Schoch, G.A.; Griffin, K.J.; Stout, C.D.; Johnson, E.F. The Structure of Human Microsomal Cytochrome P450 3A4 Determined by X-ray Crystallography to 2.05-A Resolution. *J. Biol. Chem.* **2004**, *279*, 38091–38094. [CrossRef]
17. Frishman, D.; Argos, P. Knowledge-based protein secondary structure assignment. *Proteins Struct. Funct. Bioinf.* **1995**, *23*, 566–579. [CrossRef]
18. Cojocaru, V.; Winn, P.J.; Wade, R.C. The ins and outs of cytochrome P450s. *Biochim. Biophys. Acta* **2007**, *1770*, 390–401. [CrossRef]
19. Hlavica, P. Key regulators in the architecture of substrate access/egress channels in mammalian cytochromes P450 governing flexibility in substrate oxyfunctionalization. *J. Inorg. Biochem.* **2023**, *241*, 112150. [CrossRef]
20. Urban, P.; Lautier, T.; Pompon, D.; Truan, G. Ligand Access Channels in Cytochrome P450 Enzymes: A Review. *Int. J. Mol. Sci.* **2018**, *19*, 1617. [CrossRef]

21. Hendrychova, T.; Berka, K.; Navratilova, V.; Anzenbacher, P.; Otyepka, M. Dynamics and hydration of the active sites of mammalian cytochromes P450 probed by molecular dynamics simulations. *Curr. Drug Metab.* **2012**, *13*, 177–189. [CrossRef]
22. Rydberg, P.; Rod, T.H.; Olsen, L.; Ryde, U. Dynamics of Water Molecules in the Active-Site Cavity of Human Cytochromes P450. *J. Phys. Chem. B* **2007**, *111*, 5445–5457. [CrossRef]
23. Jorgensen, W.L.; Chandrasekhar, J.; Madura, J.D.; Impey, R.W.; Klein, M.L. Comparison of simple potential functions for simulating liquid water. *J. Chem. Phys.* **1983**, *79*, 926–935. [CrossRef]
24. Loida, P.J.; Sligar, S.G. Molecular recognition in cytochrome P-450: Mechanism for the control of uncoupling reactions. *Biochemistry* **1993**, *32*, 11530–11538. [CrossRef]
25. Meng, S.; Ji, Y.; Liu, L.; Davari, M.D.; Schwaneberg, U. Modulating the Coupling Efficiency of P450 BM3 by Controlling Water Diffusion through Access Tunnel Engineering. *ChemSusChem* **2022**, *15*, e202102434. [CrossRef]
26. Fischer, A.; Smiesko, M. Spontaneous Ligand Access Events to Membrane-Bound Cytochrome P450 2D6 Sampled at Atomic Resolution. *Sci. Rep.* **2019**, *9*, 16411. [CrossRef]
27. Guengerich, F.P.; Wilkey, C.J.; Phan, T.T.N. Human cytochrome P450 enzymes bind drugs and other substrates mainly through conformational-selection modes. *J. Biol. Chem.* **2019**, *294*, 10928–10941. [CrossRef]
28. MacKerell, A.D., Jr.; Bashford, D.; Bellott, M.; Dunbrack, R.L.; Evanseck, J.D.; Field, M.J.; Fischer, S.; Gao, J.; Guo, H.; Ha, S.; et al. All-atom empirical potential for molecular modeling and dynamics studies of proteins. *J. Phys. Chem. B* **1998**, *102*, 3586–3616. [CrossRef]
29. MacKerell, A.D., Jr.; Feig, M.; Brooks, C.L., III. Improved treatment of the protein backbone in empirical force fields. *J. Am. Chem. Soc.* **2004**, *126*, 698–699. [CrossRef]
30. Best, R.B.; Zhu, X.; Shim, J.; Lopes, P.E.; Mittal, J.; Feig, M.; Mackerell, A.D., Jr. Optimization of the additive CHARMM all-atom protein force field targeting improved sampling of the backbone phi, psi and side-chain chi(1) and chi(2) dihedral angles. *J. Chem. Theory Comput.* **2012**, *8*, 3257–3273. [CrossRef]
31. Huang, J.; Rauscher, S.; Nawrocki, G.; Ran, T.; Feig, M.; de Groot, B.L.; Grubmuller, H.; MacKerell, A.D., Jr. CHARMM36m: An improved force field for folded and intrinsically disordered proteins. *Nat. Methods* **2017**, *14*, 71–73. [CrossRef]
32. Vanommeslaeghe, K.; MacKerell, A.D., Jr. CHARMM additive and polarizable force fields for biophysics and computer-aided drug design. *Biochim. Biophys. Acta* **2015**, *1850*, 861–871. [CrossRef]
33. Mark, P.; Nilsson, L. Structure and dynamics of the TIP3P, SPC, and SPC/E water models at 298 K. *J. Phys. Chem. A* **2001**, *105*, 9954–9960. [CrossRef]
34. Jing, Z.; Liu, C.; Cheng, S.Y.; Qi, R.; Walker, B.D.; Piquemal, J.P.; Ren, P. Polarizable Force Fields for Biomolecular Simulations: Recent Advances and Applications. *Annu. Rev. Biophys.* **2019**, *48*, 371–394. [CrossRef]
35. Srejber, M.; Navratilova, V.; Paloncyova, M.; Bazgier, V.; Berka, K.; Anzenbacher, P.; Otyepka, M. Membrane-attached mammalian cytochromes P450: An overview of the membrane's effects on structure, drug binding, and interactions with redox partners. *J. Inorg. Biochem.* **2018**, *183*, 117–136. [CrossRef]
36. Fischer, A.; Don, C.G.; Smieško, M. Molecular Dynamics Simulations Reveal Structural Differences among Allelic Variants of Membrane-Anchored Cytochrome P450 2D6. *J. Chem. Inf. Model.* **2018**, *58*, 1962–1975. [CrossRef]
37. Mustafa, G.; Nandekar, P.P.; Bruce, N.J.; Wade, R.C. Differing Membrane Interactions of Two Highly Similar Drug-Metabolizing Cytochrome P450 Isoforms: CYP 2C9 and CYP 2C19. *Int. J. Mol. Sci.* **2019**, *20*, 4328. [CrossRef]
38. Sansen, S.; Yano, J.K.; Reynald, R.L.; Schoch, G.A.; Griffin, K.J.; Stout, C.D.; Johnson, E.F. Adaptations for the Oxidation of Polycyclic Aromatic Hydrocarbons Exhibited by the Structure of Human P450 1A2. *J. Biol. Chem.* **2007**, *282*, 14348–14355. [CrossRef]
39. Rowland, P.; Blaney, F.E.; Smyth, M.G.; Jones, J.J.; Leydon, V.R.; Oxbrow, A.K.; Lewis, C.J.; Tennant, M.G.; Modi, S.; Eggleston, D.S.; et al. Crystal structure of human cytochrome P450 2D6. *J. Biol. Chem.* **2006**, *281*, 7614–7622. [CrossRef]
40. Berman, H.M.; Westbrook, J.; Feng, Z.; Gilliland, G.; Bhat, T.N.; Weissig, H.; Shindyalov, I.N.; Bourne, P.E. The Protein Data Bank. *Nucleic Acids Res.* **2000**, *28*, 235–242. [CrossRef]
41. Williams, P.A.; Cosme, J.; Vinković, D.M.; Ward, A.; Angove, H.C.; Day, P.J.; Vonrhein, C.; Tickle, I.J.; Jhoti, H. Crystal Structures of Human Cytochrome P450 3A4 Bound to Metyrapone and Progesterone. *Science* **2004**, *305*, 683–686. [CrossRef]
42. Sevrioukova, I.F.; Poulos, T.L. Pyridine-Substituted Desoxyritonavir Is a More Potent Inhibitor of Cytochrome P450 3A4 than Ritonavir. *J. Med. Chem.* **2013**, *56*, 3733–3741. [CrossRef]
43. Word, J.M.; Lovell, S.C.; Richardson, J.S.; Richardson, D.C. Asparagine and glutamine: Using hydrogen atom contacts in the choice of side-chain amide orientation. *J. Mol. Biol.* **1999**, *285*, 1735–1747. [CrossRef]
44. Leaver-Fay, A.; Tyka, M.; Lewis, S.M.; Lange, O.F.; Thompson, J.; Jacak, R.; Kaufman, K.; Renfrew, P.D.; Smith, C.A.; Sheffler, W.; et al. ROSETTA3: An object-oriented software suite for the simulation and design of macromolecules. *Methods Enzym.* **2011**, *487*, 545–574. [CrossRef]
45. Huang, P.S.; Ban, Y.E.; Richter, F.; Andre, I.; Vernon, R.; Schief, W.R.; Baker, D. RosettaRemodel: A generalized framework for flexible backbone protein design. *PLoS ONE* **2011**, *6*, e24109. [CrossRef]
46. Brooks, B.R.; Brooks, C.L., 3rd; MacKerell, A.D., Jr.; Nilsson, L.; Petrella, R.J.; Roux, B.; Won, Y.; Archontis, G.; Bartels, C.; Boresch, S.; et al. CHARMM: The biomolecular simulation program. *J. Comput. Chem.* **2009**, *30*, 1545–1614. [CrossRef]

47. Eastman, P.; Swails, J.; Chodera, J.D.; McGibbon, R.T.; Zhao, Y.; Beauchamp, K.A.; Wang, L.P.; Simmonett, A.C.; Harrigan, M.P.; Stern, C.D.; et al. OpenMM 7: Rapid development of high performance algorithms for molecular dynamics. *PLoS Comput. Biol.* **2017**, *13*, e1005659. [CrossRef]
48. Allen, M.P.; Tildesley, D.J. *Computer Simulation of Liquids: Second Edition*; Oxford University Press: Oxford, UK, 2017; p. 640.
49. Darden, T.; York, D.; Pedersen, L. Particle mesh Ewald: An N•log(N) method for Ewald sums in large systems. *J. Chem. Phys.* **1993**, *98*, 10089–10092. [CrossRef]
50. Wennberg, C.L.; Murtola, T.; Páll, S.; Abraham, M.J.; Hess, B.; Lindahl, E. Direct-Space Corrections Enable Fast and Accurate Lorentz–Berthelot Combination Rule Lennard-Jones Lattice Summation. *J. Chem. Theory Comput.* **2015**, *11*, 5737–5746. [CrossRef]
51. Kubo, R.; Toda, M.; Hashitume, N. *Statistical Physics II: Nonequilibrium Statistical Mechanics*, 2nd ed.; Springer: New York, NY, USA, 1991.
52. Åqvist, J.; Wennerström, P.; Nervall, M.; Bjelic, S.; Brandsdal, B.O. Molecular dynamics simulations of water and biomolecules with a Monte Carlo constant pressure algorithm. *Chem. Phys. Lett.* **2004**, *384*, 288–294. [CrossRef]
53. Chow, K.-H.; Ferguson, D.M. Isothermal-isobaric molecular dynamics simulations with Monte Carlo volume sampling. *Comput. Phys. Commun.* **1995**, *91*, 283–289. [CrossRef]
54. Ryckaert, J.P.; Ciccotti, G.; Berendsen, H.J.C. Numerical integration of Cartesian equations of motion of a system with constraints: Molecular dynamics of n-alkanes. *J. Comput. Phys.* **1977**, *23*, 327–341. [CrossRef]
55. Andersen, H.C. RATTLE: A "velocity" version of the SHAKE algorithm for molecular dynamics calculations. *J. Comput. Phys.* **1983**, *52*, 24–34. [CrossRef]
56. Miyamoto, S.; Kollman, P.A. Settle: An analytical version of the SHAKE and RATTLE algorithm for rigid water models. *J. Comput. Chem.* **1992**, *13*, 952–962. [CrossRef]
57. Humphrey, W.; Dalke, A.; Schulten, K. VMD: Visual molecular dynamics. *J. Mol. Graph.* **1996**, *14*, 33–38. [CrossRef]
58. Durell, S.R.; Brooks, B.R.; Ben-Naim, A. Solvent-induced forces between two hydrophilic groups. *J. Phys. Chem.* **1994**, *98*, 2198–2202. [CrossRef]
59. Beglov, D.; Roux, B. Finite representation of an infinite bulk system: Solvent boundary potential for computer simulations. *J. Chem. Phys.* **1994**, *100*, 9050–9063. [CrossRef]
60. Luo, Y.; Roux, B. Simulation of Osmotic Pressure in Concentrated Aqueous Salt Solutions. *J. Phys. Chem. Lett.* **2010**, *1*, 183–189. [CrossRef]
61. Won, Y. Force field for monovalent, divalent, and trivalent cations developed under the solvent boundary potential. *J. Phys. Chem. A* **2012**, *116*, 11763–11767. [CrossRef]

molecules

MDPI

Article

In Silico and In Vitro Identification of 1,8-Dihydroxy-4,5-dinitroanthraquinone as a New Antibacterial Agent against *Staphylococcus aureus* and *Enterococcus faecalis*

Juliana Amorim, Viviana Vásquez, Andrea Cabrera, Maritza Martínez and Juan Carpio *

Unidad de Salud y Bienestar, Facultad de Bioquímica y Farmacia, Universidad Católica de Cuenca, Av. Las Américas, Cuenca 010105, Ecuador
* Correspondence: juanbiomarce@gmail.com

Abstract: Increasing rates of bacterial resistance to antibiotics are a growing concern worldwide. The search for potential new antibiotics has included several natural products such as anthraquinones. However, comparatively less attention has been given to anthraquinones that exhibit functional groups that are uncommon in nature. In this work, 114 anthraquinones were evaluated using in silico methods to identify inhibitors of the enzyme phosphopantetheine adenylyltransferase (PPAT) of *Staphylococcus aureus*, *Enterococcus faecalis*, and *Escherichia coli*. Virtual screenings based on molecular docking and the pharmacophore model, molecular dynamics simulations, and free energy calculations pointed to 1,8-dihydroxy-4,5-dinitroanthraquinone (DHDNA) as the most promising inhibitor. In addition, these analyses highlighted the contribution of the nitro group to the affinity of this anthraquinone for the nucleotide-binding site of PPAT. Furthermore, DHDNA was active in vitro towards Gram-positive bacteria with minimum inhibitory concentration (MIC) values of 31.25 μg/mL for *S. aureus* and 62.5 μg/mL for *E. faecalis* against both antibiotic-resistant isolates and reference strains but was ineffective against *E. coli*. Experiments on kill-time kinetics indicated that, at the tested concentrations, DHDNA produced bacteriostatic effects on both Gram-positive bacteria. Overall, our results present DHDNA as a potential PPAT inhibitor, showing antibacterial activity against antibiotic-resistant isolates of *S. aureus* and *E. faecalis*, findings that point to nitro groups as key to explaining these results.

Keywords: antibacterial activity; nitrated anthraquinone; Gram-positive; pharmacophore; molecular docking; molecular dynamics

1. Introduction

The widespread threat of bacterial resistance leading to a decreased availability of therapeutic resources is a growing global concern [1]. The indiscriminate and unguided use of antibiotics in the treatment of bacterial infections has played a major role in accelerating this scenario [2,3]. Among the bacterial species exhibiting multidrug resistance are *Staphylococcus aureus*, *Enterococcus faecalis*, and *Escherichia coli* [4]. Although these bacteria are part of the normal human microbiota, under certain circumstances, they can cause a wide range of diseases. For this reason, there is a continuous search for new molecules with therapeutic potential [5]. The main mechanisms of bacterial resistance include mutations at the target sites that reduce affinity for the drug [6], the expression of efflux pumps that actively remove antibiotics from the cytoplasm [7], alterations in the membrane and cell wall that result in the reduced entry of antibiotics [8], and additionally, the expression of antibiotic-inactivating enzymes [9].

Phosphopantetheine adenylyltransferase (PPAT) is a key enzyme that, in bacteria, catalyzes the penultimate reaction in the synthesis of coenzyme A (CoA) [10]. This step consists of the transfer of an AMP moiety from ATP to 4′-phosphopantetheine, yielding dephospho-CoA and inorganic pyrophosphate. Due to the essential role of CoA in various cellular processes, such as amino acid metabolism and the biosynthesis of sterols, as well as the TCA cycle and fatty acid metabolism, this molecule is of central importance in bacterial metabolism [11–13]. Considering all these roles, the inhibition of this pathway is a very attractive strategy for antibacterial drug development [13]. Structurally, PPAT has three subsites, S1, S2, and S3, at the binding site, which accept and orient both substrates for catalysis. ATP binds at the S1 and S2 sites, while 4′-phosphopantetheine binds at S2 and S3. Previous studies have shown that some cycloalkyl pyrimidines inhibit the PPAT of *S. aureus*, acting as competitive and mixed inhibitors of 4′-phosphopantetheine and ATP, respectively [14].

Anthraquinones are a diverse group of molecules found mainly in plants and fungi but also present in some bacteria and insects [15]. Structurally, these molecules consist of three linearly fused six-membered rings resulting in a planar structure, with two ketone groups located at positions 9 and 10 of the central backbone ring [16]. Naturally occurring anthraquinones usually have functional groups such as hydroxyl, methyl, carboxyl, and methoxyl [17]. Apart from their use as natural colorants [18], anthraquinones have been continuously studied for their numerous biological effects. Among their properties are laxative [19], anti-inflammatory [20], anti-arthritic [21], anticancer [22], antiviral [23], and antifungal [24], as well as antibacterial potential [25–27]. The antibacterial effects of anthraquinones have been attributed to various mechanisms of action. These include the disruption of the bacterial cell membrane [28], inhibition of critical enzymes of bacterial metabolism [29], disturbance of cytokinesis, and alteration of DNA conformation [28,30]. In addition, due to their affinity for the nucleotide-binding sites (NBS) of dehydrogenases, kinases, and ATPases, some anthraquinones have been used in certain enzyme purification protocols [31].

To improve and expand the range of bioactivities of naturally occurring anthraquinones, efforts have been made for the synthesis of derivatives with functional groups not usually found in those extracted from natural sources. For instance, nitro derivatives such as 1,8-dihydroxy-4-nitro-anthraquinone have demonstrated a higher inhibitory activity against casein kinase-2 when compared to 1,8-dihydroxyanthraquinone [32]. Furthermore, the chlorinated derivative of naturally occurring emodin (3-methyl-1,6,8-trihydroxyanthraquinone) exhibits greater activity against some Gram-positive bacteria than the parent compound [28]. In the present study, we conducted in silico evaluations of 114 commercially available anthraquinones in search of potential PPAT inhibitors, followed by in vitro experiments to assess the antibacterial activity of the most promising anthraquinone against *S. aureus*, *E. faecalis*, and *E. coli*.

2. Results

2.1. Virtual Screenings Based on Molecular Docking and Pharmacophore Model

Based on the known ability of certain anthraquinones to bind to the nucleotide-binding site (NBS) of some enzymes [31] the potential of 114 molecules to inhibit PPAT was evaluated, given that it is a validated pharmacological target [14]. Figure 1A shows the comparison of the three-dimensional structures of the PPATs of *S. aureus* (*Sa*PPAT; PDB: 4NAU), *E. faecalis* (*Ef*PPAT; PDB: 3ND6), and *E. coli* (*Ec*PPAT; PDB: 6CCO) highlighting their different subsites. Importantly, in contrast to *Sa*PPAT and *Ef*PPAT, *Ec*PPAT was not co-crystallized with any ligand in the NBS. The virtual screening based on molecular docking identified 1,8-dihydroxy-4,5-dinitroanthraquinone (DHDNA) as the ligand with the best binding affinity for Gram-positive PPATs, (Supplementary Table S1). On the other hand, DHDNA ranked 52nd for *Ec*PPAT, despite having a quantitatively similar affinity to that for *Sa*PPAT and *Ef*PPAT. Interestingly, the derivative without nitro groups, 1,8-dihydroxyanthraquinone, exhibited a higher affinity for *Ec*PPAT, ranking 3rd, but, in complex with *Sa*PPAT and *Ef*PPAT, ranked 47th, and 78th, respectively. However, the derivative without hydroxyl groups, 1,8-dinitroanthraquinone, ranked 54th for *Sa*PPAT and 56th for both *Ef*PPAT and *Ec*PPAT. These results show that the presence of both the hydroxyl and nitro groups on the anthraquinone backbone is essential for the higher affinity of

DHDNA for the PPAT of both Gram-positive bacteria, but for *Ec*PPAT, the hydroxyl groups are the main contributors to the affinity. Furthermore, when analyzing the effect of the substitution of hydroxyl groups with chlorine, exemplified by 1,8-dichloroanthraquinone, the affinity for PPAT of the three species was reduced, ranking 114th, 89th, and 47th for *Sa*PPAT, *Ef*PPAT, and *Ec*PPAT, respectively.

A possible explanation for the differences in screening results obtained using the PPAT of Gram-positive bacteria, compared to those obtained with the *E. coli* enzyme, may lie in the structure of its active sites as a consequence of the absence of a co-crystallized ligand at the NBS of *Ec*PPAT. Despite the high homology in NBS sequences across bacterial species, there are marked differences in the conformation of critical PPAT residues in the ATP-bound state compared to the unbound state [13].

Since molecular docking is very sensitive to even small variations in the structure of the active site [33], the dataset was evaluated using pharmacophore-based virtual screening as an orthogonal method to identify the best ligands. Due to the limited number of competitive inhibitors of PPAT identified, pharmacophoric features were inferred from the analysis of the interactions of the co-crystallized ATP analog (AGS) at the NBS of 4NAU, Figure 1B.

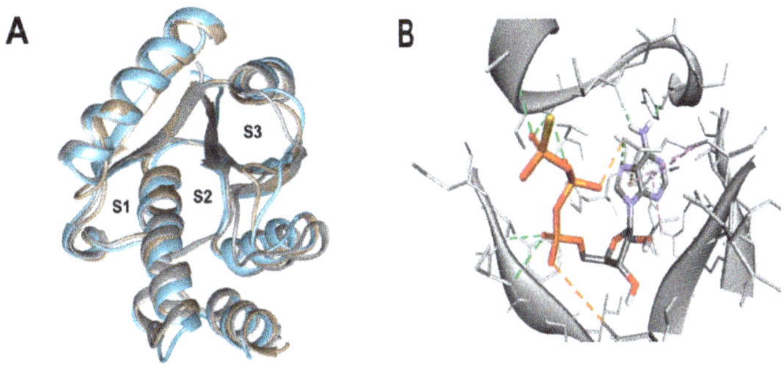

Figure 1. Three-dimensional representation of PPAT of *S. aureus*, *E. faecalis*, and *E. coli*, and 3D representation of *Sa*PPAT complexed with AGS. (**A**) Three-dimensional representation of the superposition of 4NAU (*Sa*PPAT, gray), 3ND6 (*Ef*PPAT, tan), and 6CCO (*Ec*PPAT, light blue) highlighting the subsites of interaction with ATP analog (S1 and S2), and with 4′-phosphopantetheine (S2 and S3). (**B**) Three-dimensional representation of 4NAU co-crystallized with AGS, highlighting the interactions at the NBS.

In addition to this criterion, the structure of the molecules in the dataset was also taken into account [34], since, unlike AGS, anthraquinones are planar and most of the 114 molecules have between 0 and 2 rotational bonds, as well as only three rings, corresponding to their backbone, Figure 2A. The query pharmacophore model included: (1) hydrogen bond donor (amino group of adenine involved in hydrogen bonds with TYR125 and ILE128), (2) aromatic ring (imidazole forming pi-stacking interactions with ARG92), and (3) hydrogen bond acceptor (oxygen atom of alpha-phosphate involved in two hydrogen bonds with SER11 and PHE12), Figures 2B and 3A.

The screening results show that, among the 114 anthraquinones, only 14 satisfied the pharmacophore model and had RMSD values < 0.8. At the same time, ATP included as the positive control had the best score (RMSD = 0.07), Supplementary Table S2. Notably, DHDNA once again ranked among the best ligands, reaching the second position (RMSD = 0.58). Furthermore, when exclusive shape constraints were set to a tolerance level of 0, DHDNA was the best ligand (RMSD = 0.43), indicating that its conformation in the active site matches the pharmacophore model without steric hindrance [35]. As shown in Figure 3B, one of its hydroxyls acts as a hydrogen bond donor, the anthraquinone core meets the pharmacophore requirement of the aromatic system, and the oxygen in the nitro group acts as a hydrogen bond acceptor. Taken together, these results support the hypothesis that DHDNA has the potential to target the NBS of PPAT.

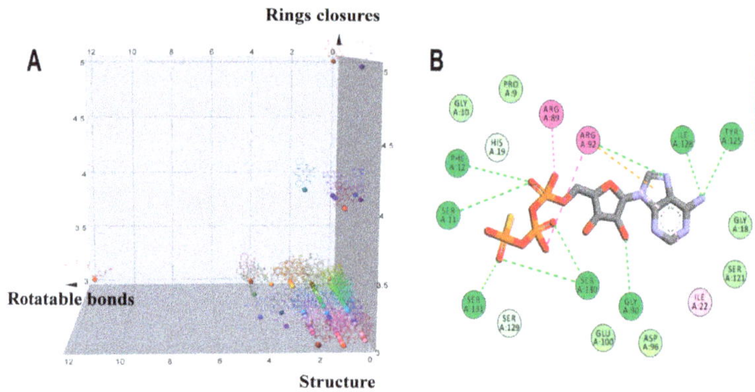

Figure 2. Analysis of the structural diversity of the database composed of 114 anthraquinones and a 2D interaction diagram of the complex 4NAU-AGS. (**A**) Three-dimensional distribution graph of the structures of 114 anthraquinones: x = number of rotatable bonds, y = number of ring closures, and z = chemical structure. (**B**) Two-dimensional diagram of the interaction of the complex 4NAU-AGS.

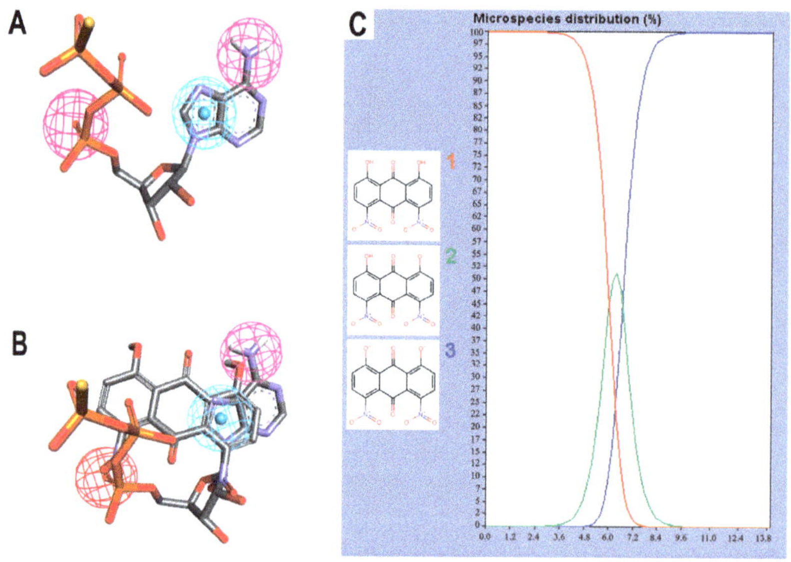

Figure 3. Pharmacophore features and species of 1,8-dihydroxy-4,5-dinitroanthraquinone (DHDNA) in the full pH range. (**A**) Features used to create the query pharmacophore model: hydrogen bond donor, aromatic system, and hydrogen bond acceptor are represented in magenta, blue and red spheres, respectively. (**B**) Overlap of AGS with DHDNA also represented in the same colors. These representations were generated with Discovery Studio using the coordinates obtained from Pharmit analyses. (**C**) Distribution of the three species of the DHDNA across the entire pH range, (1) protonated, (2) semi-protonated, and (3) deprotonated species.

Considering the strong effect of the nitro group on the electronic properties of organic molecules [32,36], it was decided to evaluate its influence on the protonation state of the two hydroxyl groups of the DHDNA structure. The results show that, at the cytoplasmic pH range from 7.2 to 7.6, the predominant species is the deprotonated (89–75.5%), followed by the semi-protonated (23–11%) and a small fraction of the protonated state (1.5–0.3%), Figure 3C. Based on these results, for the following analyses, the three states, i.e., protonated (pDHDNA), semi-protonated (sDHDNA), and deprotonated (dDHDNA), were included because, despite the lower abundance of sDHDNA and pDHDNA, their involvement in potential biological effects could not be ruled out.

To better understand the DHDNA interactions at the active site of the three PPATs, the interaction diagrams of the molecular docking results were analyzed. The results of the three species of DHDNA complexed with *Sa*PPAT (Figure 4A–C) show that their substituent groups (-OH, -O$^-$, and -NO_2) participated in interactions at the NBS, but dDHDNA established a lower number of hydrogen bonds. Notably, all the DHDNA states interact with critical residues, such as HYS19, which stabilizes ATP to nucleophilic attack at the α-phosphate group; ARG92, which is involved in the stabilization of the β-phosphate of the nucleotide; and the conserved SER130, which is part of a three-serine stretch located at the floor of subsite S1 [13].

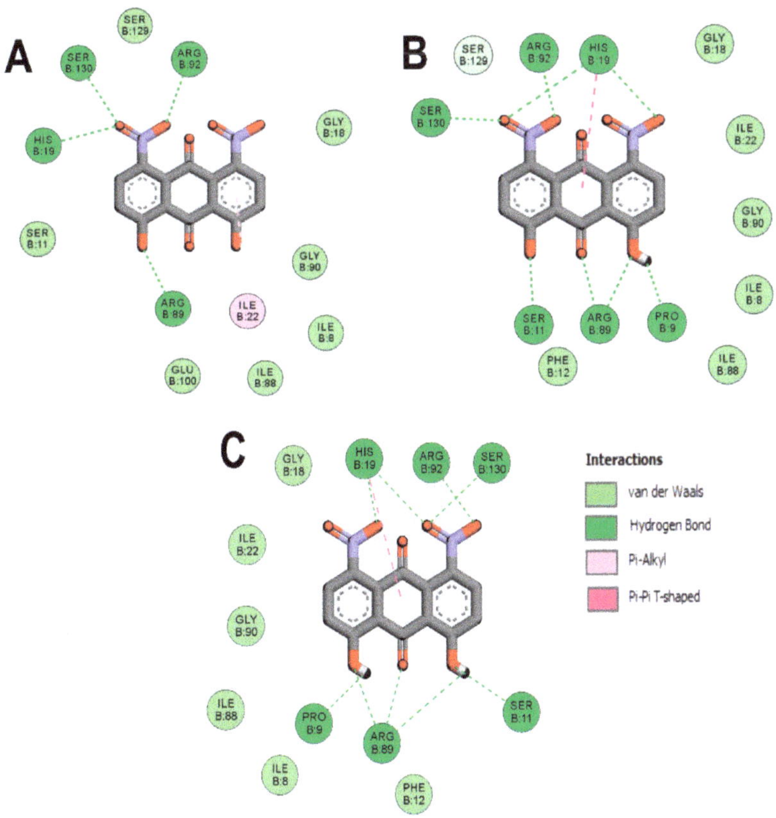

Figure 4. Analysis of the docked poses of the three species of DHDNA in a complex with 4NAU. (**A**) Two-dimensional interaction diagrams of 4NAU complexed with dDHDNA, (**B**) sDHDNA, and (**C**) pDHDNA.

On the other hand, the interactions of *Ef*PPAT with the three DHDNA species were less conserved, with only ARG92 and VAL128 interacting with all of them, Figure 5A–C. In addition, dDHDNA established more interactions with this target than with *Sa*PPAT, which is reflected in a binding affinity close to that of sDHDNA and pDHDNA. These results support the hypothesis that one of the main contributions of nitro groups to the binding affinities of DHDNA to these targets could be to favor the formation of interactions with polar amino acids in the active site.

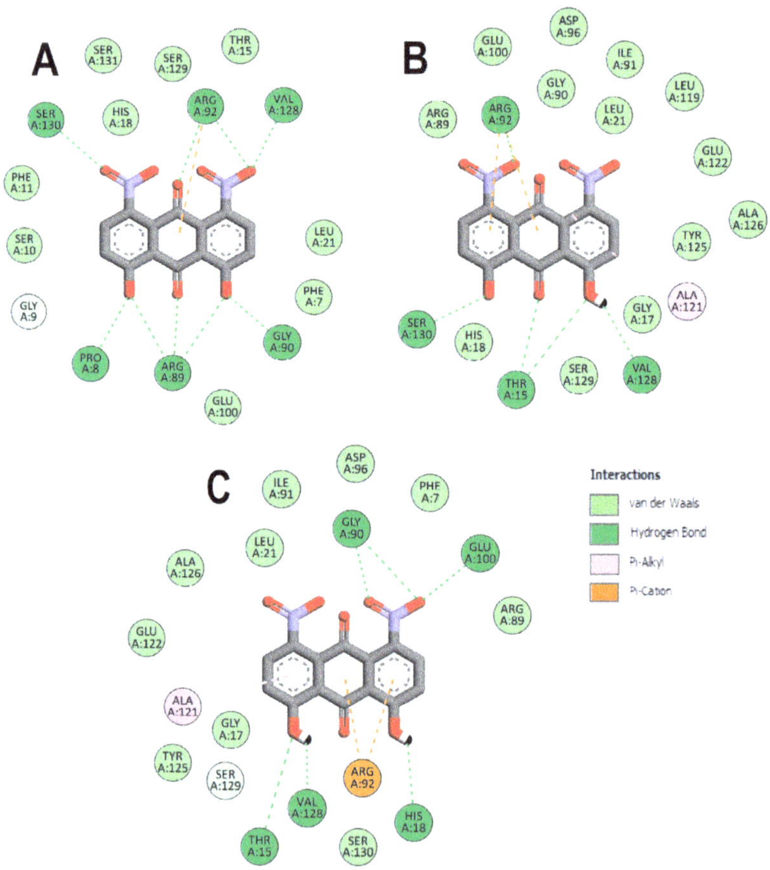

Figure 5. Analysis of the docked poses of the three states of DHDNA in complex with 3ND6. (**A**) Two-dimensional interaction diagrams of 3ND6 complexed with dDHDNA, (**B**) sDHDNA, and (**C**) pDHDNA.

In the case of *Ec*PPAT, the only common hydrogen bond formed with the three DHDNA species involved ARG91, while TYR7, THR15, SER129, and SER130 were common to the complexes with the dDHDNA and sDHDNA states. At the same time, ARG88 only established bonds with the sDHDNA form, while THR10, PHE11, and SER128 interacted exclusively with the pDHDNA species, Figure 6A–C.

Considering that the affinities of the three species to the respective enzymes are close, collectively these results suggest that the effects of nitro groups on the deprotonation of hydroxyl groups, giving rise to negative charges, would not be the cause of the higher affinity of DHDNA for NBS compared to that of the other ligands analyzed.

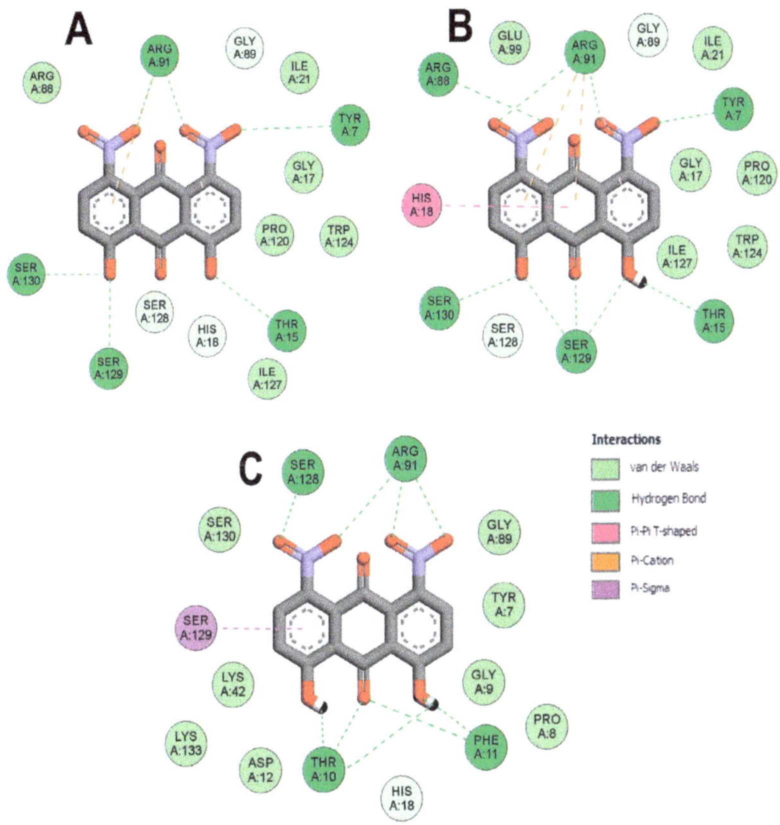

Figure 6. Analysis of the docked poses of the three states of DHDNA in complex with 6CCO. (**A**) Two-dimensional interaction diagrams of 6CCO complexed with dDHDNA, (**B**) sDHDNA, and (**C**) pDHDNA.

2.2. *Molecular Dynamic Simulations and Total Binding Free Energy Calculations*

Molecular dynamic simulations were performed to evaluate the stability of the complexes formed between the three DHDNA species with each PPAT enzyme. Figure 7A shows that the profile of complexes with *Sa*PPAT was more stable compared to that with *Ef*PPAT (Figure 8A), but less stable compared to *Ec*PPAT (Figure 9A), throughout the time analyses. The complex with pDHDNA increased the RMSD during the first 2 ns until it reached 0.5 nm, a value maintained until the end of the run. At the same time, the *Sa*PPAT-dDHDNA complex showed, during the first 21 ns, an RMSD of a maximum of 0.25 nm, followed by an increase to a value of 0.4 nm, which was maintained until the end of the analysis time. In the case of the complex with sDHDNA, it presented an RMSD of less than 0.25 nm for most of the analysis time. Moreover, in the last 40 ns, this profile was similar to that of the co-crystallized ligand, which corroborates the higher stability of the complex with the semi-protonated state compared to those formed with the other species. Furthermore, as the negative values of the total binding energy calculated with MMPBSA suggest, the three DHDNA species, as well as the co-crystallized ligand (AGS), remain in the enzyme throughout the analysis time, Figure 7B.

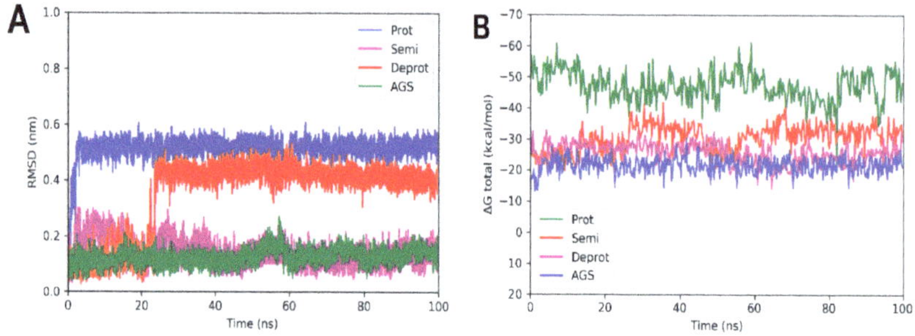

Figure 7. RMSD analyses and calculations of the binding free energy of 4NAU complexed with three states of DHDNA and with the co-crystallized ligand (AGS). (**A**) RMSD values. (**B**) Total ΔG energies.

On the other hand, the analyses of the complexes with *Ef*PPAT reveal RMSD values with greater fluctuations during most of the runs, Figure 8A. Among them, the complex with the semi-protonated form exhibited relatively better stability, showing an RMSD below 0.38 nm during most of the analysis time. However, despite fluctuations, even the less stable complex formed with the deprotonated state also did not leave the enzyme throughout the run, apparently, as suggested in its total binding energy plot (negative values), Figure 8B. It is noteworthy that, in the complexes with both enzymes of Gram-positive bacteria, there are no marked differences between the total binding energies, results that agree with the docking analyses.

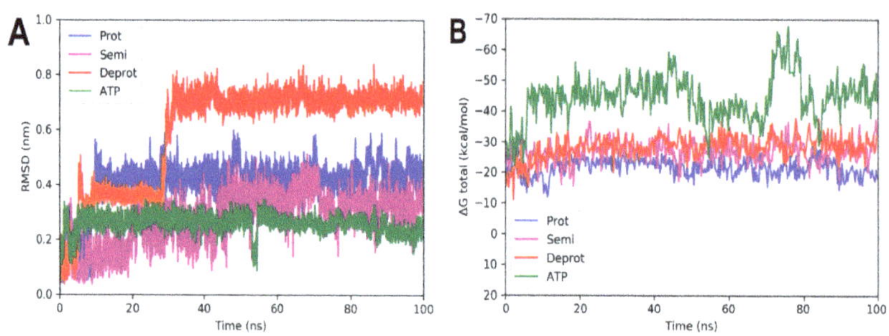

Figure 8. RMSD and free energy binding calculation analyses of 3ND6 complexed with three states of DHDNA and with the co-crystallized ligand (ATP). (**A**) RMSD values. (**B**) Total ΔG energies.

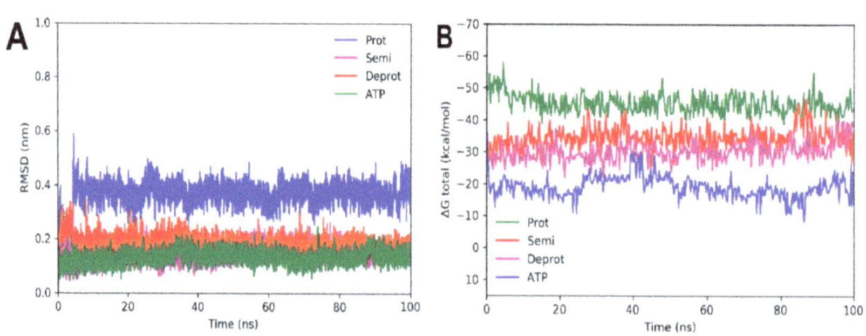

Figure 9. RMSD and free energy binding calculation analyses of 6CCO complexed with three states of DHDNA and with docked ATP. (**A**) RMSD values. (**B**) Total ΔG energies.

For the complexes involving *Ec*PPAT and the three DHDHA states, the smallest variations in the RMSD values obtained were observed. The complex with the protonated form increased by 0.4 nm in the first 3 ns of the simulation, while the deprotonated and semi-protonated forms maintained an average of 0.18 and 0.17 nm, respectively, Figure 9A. In fact, the values of the latter two forms are very similar to those of the docked ATP used as control. In the same way as the other crystals, the total energy values remained practically constant throughout the simulation time, suggesting that the complexes remained integrated over time, Figure 9B.

2.3. Decomposition of Binding Free Energy Analysis

To further analyze the contributions of the different energies to the affinity of DHDNA for the three PPATs, binding energy decompositions were performed from calculations of the total binding free energy of the last 50 ns of each of the runs. The total ΔG for *Sa*PPAT complexed with pDHDNA, sDHDNA, and dDHDNA, were, respectively, −22.5, −24.3, and −31.0 kcal/mol (Supplementary Table S3). Additionally, among *Ef*PPAT and, respectively, pDHDNA, sDHDNA, and dDHDNA, the total ΔG were −21.1, −25.0, and −29.7 kcal/mol (Supplementary Table S4). At the same time, for the different states of DHDNA interacting with *Ec*PPAT, the authors obtained values of −21.8, −29.9, and −34.2 kcal/mol, respectively, for pDHDNA, sDHDNA, and dDHDNA (Supplementary Table S5).

In all the complexes studied, it was observed that the protonated state of DHDNA presents a higher contribution from Van der Waals interactions (ΔE, vdw), while the contributions from electrostatic interactions (ΔE, ele) are higher for the semi- and deprotonated forms. Although the latter component increases as a consequence of the deprotonation of one or two of the hydroxyl groups of DHDNA, respectively, the total energies of the three species are very close.

Furthermore, when analyzing the total ΔG of the complexes between the co-crystallized ligand of *Sa*PPAT (AGS), *Ef*PPAT (ATP), ATP docked to *Ec*PPAT, and the three DHDNA species, the values are markedly higher in the complexes involving nucleotides. This is an expected result, given that nucleotides not only occupy the pocket of adenosine at the NBS but also project at the site destined to interact with the phosphates of endogenous ATP. In fact, one possibility to further improve the binding energy of DHDNA with this enzyme would be to introduce modifications in its structure in an attempt to more closely resemble the interactions formed by phosphates of the endogenous ligand.

2.4. Pharmacokinetic and Target Fishing Predictions

Due to the limited information on biological assays with DHDNA, in silico analyses were conducted to obtain preliminary information on its pharmacokinetic profile and to identify potential targets in humans. As shown in Supplementary Table S6, predictions performed with SwissADME reveal that this anthraquinone has low gastrointestinal absorption, no potential to cross the blood–brain barrier (BBB), and could be a P-gp substrate. Considering these results, and in view of future in vivo assays, it may be necessary to develop a suitable formulation for the infection model selected for these experiments. In relation to metabolism, among the five cytochromes included in the predictive analyses, DHDNA would only be able to inhibit CYP2C9, which should be taken into account in possible future studies involving drugs metabolized by this enzyme.

Regarding the biological activity of DHDNA, it has been reported as a potent inhibitor of the dengue NS2B-NS3 viral protease [37] but, to our knowledge, there are no previous reports on their toxicological evaluation. Considering that anthraquinones from natural sources, such as rhein or emodin, can interact with human enzymes [16], target fishing analyses were conducted to identify possible human targets for DHDNA. The results in Supplementary Tables S7–S9 show that, compared to emodin and rhein, DHDNA has a lower probability of interacting with human targets, which may suggest a lower probability of causing off-target effects. Finally, the prediction of cytotoxicity to NIH/3T3 cells (mouse embryonic fibroblast) in Supplementary Table S10 indicates that DHDNA is not cytotoxic.

2.5. In Vitro Evaluation of the Antibacterial Activity of DHDNA

To evaluate the effect of DHDNA against *S. aureus*, *E. faecalis*, and *E. coli*, nine isolates, as well as reference strains, of each species were exposed to this compound using the agar macrodilution method. Since most of the bacteria we isolated showed resistance to ciprofloxacin, the nine isolates of each species we selected showed resistance to this antibiotic, Figure 10A. The results show that DHDNA at a concentration of 125 µg/mL completely inhibited the growth of *S. aureus* and *E. faecalis* (in both isolates and reference strains), Figure 10B, compared to the growth of both bacteria in the control medium for *S. aureus*, Figure 10C, and for *E. faecalis*, Figure 10D.

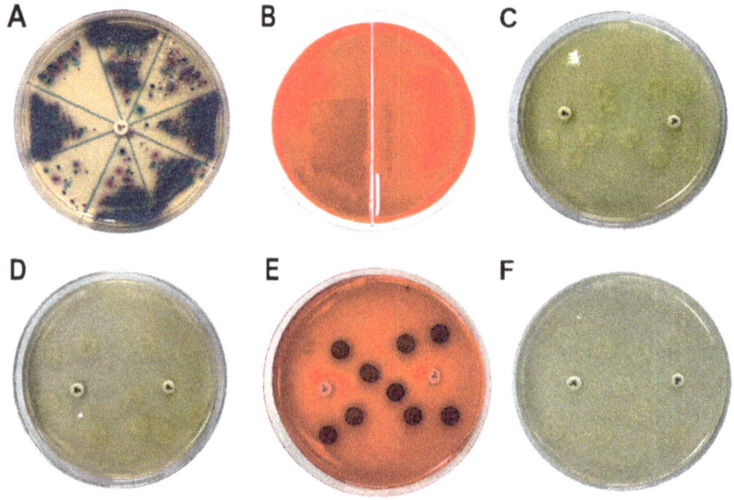

Figure 10. Samples streaked in chromogenic agar to isolate antibiotic-resistant *S. aureus*, *E. faecalis*, and *E. coli* and the effects of DHDNA on bacterial growth. (**A**) Chromogenic agar plate with ciprofloxacin disc (5 µg) to identify and isolate antibiotic-resistant bacteria. (**B**) *S. aureus* (left) and *E. faecalis* (right) exposed to 125 µg/mL of DHDNA. (**C**) *S. aureus* on agar plates with 1% DMSO in the proximity of ciprofloxacin discs to confirm antibiotic resistance. (**D**) *E. faecalis* in the same anterior conditions. (**E**) Colonies of *E. coli* exposed to 125 µg/mL of DHDNA and (**F**) on agar plates with 1% DMSO. DHDNA causes the red color of the agar plate. The images are representative of at least three independent experiments.

It is noteworthy that, when we exposed these bacteria to the anthraquinone without nitro groups, 1,8-dihydroxyanthraquinone, no growth inhibition was detected, demonstrating that the introduction of nitro groups is responsible for the antibacterial effect of DHDNA, Table 1. A previous study also reported the ineffectiveness of 1,8-dihydroxyanthraquinone against methicillin-resistant *S. aureus* [38]. Similarly, we observed that *S. aureus* and *E. faecalis* exposed to 1,8-dichloroanthraquinone did not show any differences concerning growth in the control medium, contrary to the enhanced antibacterial effect against *S. aureus* reported by the introduction of a chlorine atom in emodin [28].

Compared to the greater attention that *S. aureus* has attracted, relatively few studies have evaluated the effect of anthraquinones on *E. faecalis*. Among these few molecules tested, 1-(2-aminoethyl)piperazinyl-9,10-dioxo-anthraquinone [39] and emodin [28] have been reported to be inactive. In this context, the antibacterial effect exhibited by DHDNA is a relevant finding that highlights the contribution of nitro groups for the search and design of new anthraquinones active against Gram-positive bacteria. Although future in vitro experiments will be necessary to confirm the inhibition of PPAT by DHDNA, it is noteworthy that both the antibacterial activity of this compound, as well as the ineffectiveness of

1,8-dihydroxyanthraquinone and 1,8-dichloroanthraquinone, are in agreement with the in silico results.

Table 1. Growth of *S. aureus*, *E. faecalis*, and *E. coli* isolates and control strains in the presence of 1,8-dihydroxy-4,5-dinitroanthraquinone, 1,8-dihydroxyanthraquinone, and 1,8-dichloroanthraquinone at the concentration of 125 μg/mL.

Results of Growth of *S. aureus*, *E. faecalis*, and *E. coli* Exposed to Selected Anthraquinones.			
	S. aureus	***E. faecalis***	***E. coli***
1,8-dihydroxy-4,5-dinitroanthraquinone 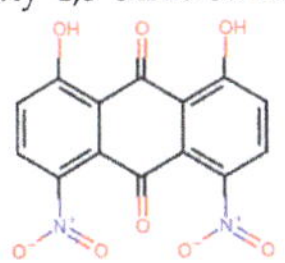	Absence	Absence	Presence
1,8-dihydroxyanthraquinone 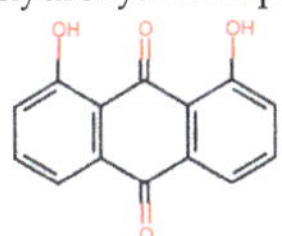	Presence	Presence	Presence
1,8-dichloroanthraquinone 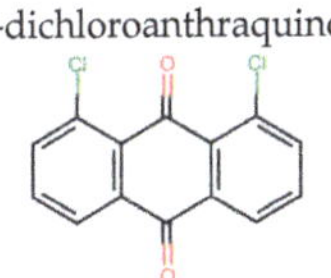	Presence	Presence	Presence

On the other hand, when DHDNA was tested in the isolated and reference strains of *E. coli* under the same conditions mentioned above, no inhibitory effect was detected, Figure 10E. Interestingly, the bacteria exposed to DHDNA adopted a dark purple color, although they maintained the same morphology as the colonies in the control medium, Figure 10F. The results of the inactivity of DNDNA are in line with previous studies that have shown null or a reduced effect of several anthraquinones on Gram-negative bacteria. For example, the absence of antibacterial activity on *E. coli* was reported of anthraquinone (without functional groups) (100 μM), 1,5-dihydroxianthraquinone (10 μM), and 1,8-dihydroxianthraquinone (10 μM) [40]. In another recent study, anthraquinone mitoxantrone was up to 20 times less potent on Gram-negative bacteria compared to its effect on Gram-positive bacteria [41].

In addition, other authors, also, have shown that certain anthraquinones with hydroxyl, methoxyl, and carboxyl groups have low or no activity towards Gram-negative bacteria [42–44]. Among the causes of the reduced sensitivity of these bacteria to various antimicrobial agents are mechanisms that prevent or reduce their intracellular accumulation [45]. In fact, it has been proposed that the antibacterial effects of anthraquinones, such as emodin, could be explained mainly by their ability to cause direct damage to the bacterial membrane [30,46], or at least as a consequence of destabilizing it to allow them access to their intracellular targets [47]. In this context, the inactivity of DHDNA against *E. coli* could indicate that it does not exert such effects on the membrane, a hypothesis that will require future experiments to verify. On the other hand, structure–activity relationship studies have revealed that the presence of a primary amino group is a structural feature that favors the entry and retention of drugs in Gram-negative bacteria [48,49]. Considering that DHDNA does not include such a group, a lower intercellular accumulation of this compound could be an additional hypothesis for its lack of effect. Following this line of reasoning, the incorporation or substitution of some group of DHDNA by amino groups would allow the testing of this hypothesis and potentially expand its antibacterial spectrum.

2.6. Evaluation of the Potential of DHDNA to Resensitize Antibiotic-Resistant Bacteria

To obtain preliminary evidence of the potential of DHDNA to recover the effect of antibiotics against resistant bacteria, one isolate of each species exhibiting resistance to a greater number of antibiotics was selected among the samples used in previous experiments. The results in Table 2 show that the presence of DHDNA, at subinhibitory concentrations, did not sensitize any of the isolates to the effects of the antibiotics tested. A previous study showed that mitoxantrone synergizes vancomycin and other antibiotics such as ciprofloxacin against resistant *E. faecalis* strains. The synergism with vancomycin was related to the ability of this anthraquinone to induce oxidative stress and DNA damage [41]. Although it has been reported that their effects on the bacterial membrane could be another mechanism for anthraquinones to induce synergism [50], further studies will be needed to determine whether DHDNA does not sensitize the tested bacteria because it does not elicit such effects.

Table 2. Evaluation of the potential of DHDNA to resensitize resistant strains of *S. aureus*, *E. faecalis*, and *E. coli* against antibiotics. Selected colonies of *S. aureus*, *E. faecalis*, and *E. coli* were exposed to DHDNA dissolved in agar plates (15.5 μg/mL, 31.125 μg/mL, and 125 μg/mL, respectively) and incubated in the presence of antibiotic discs. The results presented were obtained from at least three independent experiments.

Effect of the Presence of Sub-MIC Concentrations of DHDNA in Bacteria Resistant to Selected Antibiotics	
S. aureus	
Ciprofloxacin (5 μg)	Resistant
Azithromycin (15 μg)	Resistant
Chloramphenicol (30 μg)	Resistant
Erythromycin (15 μg)	Resistant
Tetracycline (30 μg)	Resistant
Trimethoprim/sulfamethoxazole (25 μg)	Resistant
E. faecalis	
Ciprofloxacin (5 μg)	Resistant
Clindamycin (2 μg)	Resistant
Cefoxitin (30 μg)	Resistant
Cefuroxime (30 μg)	Resistant
Tetracycline (30 μg)	Resistant
Trimethoprim/sulfamethoxazole (25 μg)	Resistant
E. coli	
Ciprofloxacin (5 μg)	Resistant
Clarithromycin (15 μg)	Resistant
Ampicillin (10 μg)	Resistant
Amoxicillin (25 μg)	Resistant
Cephalexin (30 μg)	Resistant
Cefuroxime (30 μg)	Resistant
Chloramphenicol (30 μg)	Resistant
Trimethoprim/sulfamethoxazole (25 μg)	Resistant

2.7. Determination of Minimum Inhibitory Concentrations

The next step was to determine the MIC of DHDNA in both reference and antibiotic-resistant Gram-positive strains. The MIC values for *S. aureus* and *E. faecalis* were, respectively, 31.125 μg/mL and 62.5 μg/mL, in both reference and isolate strains (Table 3). Comparing the activity of DHDNA with that of the previously reported chlorinated emodin [28], the latter has a higher activity against *S. aureus* (MIC = 4 μg/mL) but lower against *E. faecalis* (MIC = 256 μg/mL). In another study, Machado et al. tested 1,3-dimethoxy-8-hydroxy-6-methylanthraquinone (MIC > 32 μg/mL), 1,3-dimethoxy-2,8-dihydroxy-6-methylanthraquinone (MIC > 64 μg/mL), and 3-propyl-pyridinium anthraquinone derivative (MIC > 32 μg/mL) against both bacteria, without detecting antibacterial effects [51].

Table 3. The minimal inhibitory concentration (MIC) of DHDNA on isolates and control strains of *S. aureus* and *E. faecalis*. The results were obtained from at least three independent experiments.

MIC (μg/mL)		
	S. aureus	***E. faecalis***
1,8-dihydroxy-4,5-dinitroanthraquinone	31.125	62.5

2.8. Time-Kill Kinetic Analysis

The effects of DHDNA on the growth kinetics of both the *S. aureus* and *E. faecalis* reference strains were studied to determine whether its effect is bacteriostatic or bactericidal. Figure 11A shows that, from eight to 24 h of exposure at concentrations of 31.125 μg/mL (1 × MIC) and 62.5 μg/mL (2 × MIC), DHDNA induced a concentration-dependent reduction trend in the number of *S. aureus* relative to control. However, this reduction was ≤3 log10 cfu/mL relative to the number of bacteria in the initial inoculum. In *E. faecalis*, after 24 h of exposure, even to a concentration of 125 μg/mL of DHDNA (2 × MIC), the number of bacteria remains essentially unchanged compared to the initial count, Figure 11B.

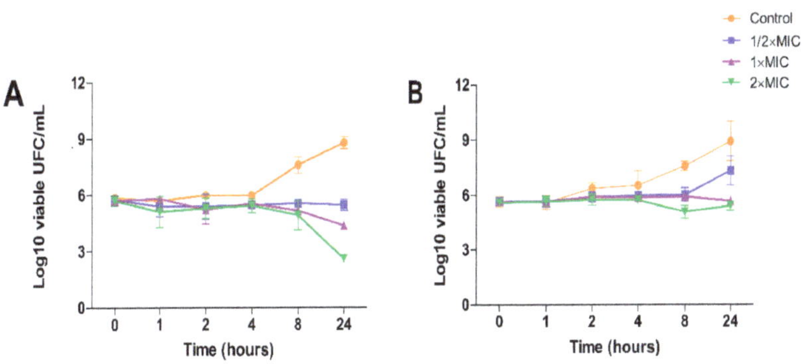

Figure 11. Time-kill kinetic assay using DHDNA at 1/2 × MIC, 1 × MIC, and 2 × MIC concentrations with exposure for 0, 1, 2, 4, 8, and 24 h. (**A**) Assay with reference strain of *S. aureus*. (**B**) Assay with reference strain of *E. faecalis*. The results are shown as the mean ± SD of at least three independent experiments.

These results suggest that DHDNA, at the tested concentrations, exerts a bacteriostatic effect in both *S. aureus* and *E. faecalis*. In line with this, a recent study reported that rhein (1,8-dihydroxyanthraquinone-3-carboxylic acid) also provoked bacteriostatic effects on *S. aureus* at concentrations of 12.5 μg/mL (1 × MIC) and 25 μg/mL (2 × MIC) [52].

It is important to note that, relative to growth on the control medium, exposure to DHDNA generated a decreasing trend in bacterial numbers only after eight hours of incubation. In this context, the delayed appearance of evidence of the effect of an agent on bacterial growth is often associated with molecules that act as inhibitors of cofactor biosynthesis [14]. Interestingly, considering that PPAT is involved in the synthesis of coenzyme A, these growth kinetic results coincide with those that would be expected in the case of inhibition of this enzyme by DHDNA.

3. Materials and Methods

3.1. Ligands and Targets Preparation for In Silico Analyses

The chemical structures of the anthraquinones and controls were downloaded from the ZINC20 database [53] in January 2023. Next, Marvin Sketch software was used to calculate the protonation state of the molecules to pH 7.4, and subsequently their 3D structures were generated with Avogadro-1.2 software [54]. Ligands were optimized by energy minimization using the MMFF94 force field with optimization of the steepest descent geometry with 500 steps, followed by the conjugate gradient algorithm with default parameters, and transformed into the MOL2 format. The analysis of the structural diversity of the anthraquinones in the database was carried out with DataWarrior V5.5.0 [55].

The affinities of selected anthraquinones for the ATP binding site of PPAT of *S. aureus*, *E. faecalis*, and *E. coli* (*Sa*PPAT, *Ef*PPAT, and *Ec*PPAT, respectively) were evaluated with molecular docking-based virtual screening. The 3D X-ray diffraction structures of *Sa*PPAT (PDB: 4NAU [14], chain B), *Ef*PPAT (PDB: 3ND6 [56], chain A), and *Ec*PPAT (PDB: 6CCO [57], chain A) were retrieved from the RCSB Protein Data Bank in January 2023. These targets were prepared using the Dock-prep module of UCSF-Chimera-1.16 [58] software applying the default parameters. The correct protonation state of certain amino acids, such as HIS-17, was inspected before docking analysis due to their critical role during catalysis [13]. Then, each structure was processed by the SPORES 1.3 tool using default parameters and saved in MOL2 format.

3.2. Molecular Docking Analyses

Molecular docking analyses were performed using the Protein-Ligand ANT System-1.2 software (PLANTS-1.2) [59]. All runs were performed with a radius of 12.5 Å, centering the coordinates on each co-crystallized ligand in the NBS, and by overlap with 4NAU for determination of this site at 6CCO. These coordinates of the x, y, and z axes were -15.5, 25.1, and 42.0 for 4NAU; -16.0, -11.0 and 31.0 for 3ND6; and -29.0, -42.0, and 51.0 for 6CCO. To ensure effective clustering, an RMSD value of 2.0 Å was established and default settings were used for all other parameters.

3.3. Pharmacophore-Based Virtual Screening

The pharmacophoric features were determined from the analysis of interactions of AGS, which is an ATP analog, co-crystallized with the A chain of 4NAU. The Pharmit server (https://pharmit.csb.pitt.edu/, (accessed on 20 November 2023)) [35] was used to perform the screening by applying the inferred features: hydrogen bond donor ($x = -4.594$, $y = -40.148$, and $z = 18.476$), aromatic system ($x = -6.7402$, $y = -41.027$, and $z = 21.0482$) from adenine moiety, and oxygen from the alpha-phosphate ($x = -13.117$, $y = -42.054$, and $z = 21.638$) as hydrogen bond acceptor. To perform the screening, the anthraquinone dataset was converted to SDF format and used as the input file to generate the respective conformers. The screenings with Pharmit were performed by applying: (1) no exclusive shape constraint, and (2) exclusive shape constraint with a tolerance level of 0.

3.4. Molecular Dynamic Simulations

All MD simulations of the complexes between the three states of DHDNA and each of the selected enzymes, as well as the complexes between the co-crystallized molecules, were performed using the software GROMACS-2021.4 [60], applying an all-atom CHARMM 36 force field [61]. The solvation water model employed for all systems was the water transferable intermolecular potential 3P (TIP3P), which was utilized within a periodically corrected cubic box, ensuring a minimum edge distance of 1.2 nm. To achieve system neutrality, Na^+ and Cl^- ions were added. The steepest descent algorithm was then employed to perform 50,000 energy minimization steps to eliminate initial steric shocks.

The equilibration process consisted of two stages. Firstly, the system was equilibrated for 500 ps at a temperature of 310 K in the NVT ensemble. Subsequently, an equilibration period of 5000 ps was conducted in the NPT ensemble at a pressure of 1 bar. The production runs were carried out for a maximum duration of 100 ns, with the coordinates saved at regular intervals of 10 ps. To ensure accurate control of pressure and temperature, the leap-frog algorithm and Berendsen coupling were employed throughout the procedures [62]. The long-range electrostatic interactions were analyzed using the particle mesh Ewald (PME) algorithm [63], while the LINCS algorithm implementation regulated the covalent bonds [64].

3.5. Binding Free Energy Calculation

We carried out total binding free energy calculations from molecular dynamic trajectories to further investigate the magnitude and types of interactions that contribute to the total energy in these complexes. The MMPBSA methodology was employed [65], utilizing a single trajectory, and calculated using gmx-MMPBSA 1.5.7 software [66]. From the MD analyses, the results of all the runs were extracted in two different ways. The first considered the entire time of each run to extract the total energy, considering 500 snapshots. Subsequently, for the energy decomposition analyses, 500 snapshots were extracted from the last 50 ns of each MD run. The determination of free energies incorporated specific parameters: inp = 1, istrng = 0.15, and indi = 2. While these parameters were utilized, the remaining parameters adhered to the recommended settings of the software.

3.6. Pharmacokinetic, Target Fishing, and Cytotoxic Predictions

The pharmacokinetic predictions were carried out using the Swiss-ADME web server (http://www.swissadme.ch/, (accessed on 20 March 2023)) [67]. The predictions of potential human targets of the selected ligands (target fishing) were performed with the Swiss Target Prediction web server (http://www.swisstargetprediction.ch/, (accessed on 20 March 2023)) [68]. The results are presented as scores ranging from 0 to 1, where the value 1 corresponds to the most likely target of the query molecule. Prediction of cytotoxicity was performed using the MouseTox web server (http://www.swisstargetprediction.ch/, (accessed on 20 March 2023)) [69], which is a tool trained to predict cytotoxic compounds for NIH/3T3 cells. All these servers were accessed in January 2023.

3.7. Material

1,8-dihydroxy-4,5-dinitroanthraquinone (97%) (CAS 81-55-0), 1,8-dihydroxyanthraquinone (97%) (CAS: 117-10-2), 1,8-dichloroanthraquinone (97%) (CAS: 82-43-9), Mueller-Hinton agar, and Mueller-Hinton broth were purchased from Sigma-Aldrich (St. Louis, MO, USA). HiCrome™ UTI Agar was purchased from Hi-media Laboratory Ltd. (Mumbai, India). All antibiotic disks were purchased from Bioanalyse (Ankara, Turkey). Control strains of *S. aureus* (ATCC 25923), *E. faecalis* (ATCC 29212), and *E. coli* (ATCC 25922) were obtained from the American Type Culture Collection (Rockville, MD, USA). Dimethyl sulfoxide (DMSO) was obtained from Merck (Rahway, NJ, USA). All reagents used were of the highest grade available.

In the experiments, the maximum concentration of anthraquinones used was 125 μg/mL. Final concentrations in the culture media were obtained by using solutions of the anthraquinones prepared in DMSO and adding them to sterile Mueller-Hinton (MH) broth or molten MH agar. The final concentration of DMSO in the culture media used in all experiments was 1%.

3.8. Isolation and Identification of Bacteria

Nine isolates each of *S. aureus*, *E. faecalis*, and *E. coli* were obtained from contaminated surfaces or wastewater from animal farms, each isolate coming from different samplings. Samples were streaked on UTI chromogenic agar for the identification of characteristic colonies of each bacterium. In addition, since we decided to evaluate the anthraquinones against resistant bacteria, antibiotic disks were incorporated into the same agar immediately after streaking and incubated at 37 °C for 16 h. Representative colonies of each bacterium were picked from the proximity of the antibiotic discs, further subcultured on selective and differential media for each species, and stained with Gram stain reagents for microscopic examination. Finally, confirmed colonies were picked, cultured in broth, and subsequently stored with glycerol (15%) at −20 °C. Reference strains from *S. aureus*, *E. faecalis*, and *E. coli* were cultured and stored in the same manner as the isolates.

3.9. Antibacterial Activity and Minimal Inhibitory Concentration Assays

The bacterial susceptibility and determination of the MIC of the anthraquinones against each bacterium were conducted following the agar macrodilution method (Clinical and Laboratory Standards Institute guidelines). For these experiments, fresh colonies were streaked on MH agar for 16 h at 37 °C. Subsequently, colonies were suspended in saline, adjusting the cell density to 1×10^8 cfu/mL. These suspensions were further diluted with a volume of saline sufficient to add on the agar surfaces (with anthraquinones and control media) a total of 25 μL of suspension containing 10^4 cfu per spot. Lastly, after incubation, the number of colonies on the plates was counted. Anthraquinones that completely inhibited bacterial growth at a concentration of 125 μg/mL were considered active.

The determination of the MIC was carried out following the protocol mentioned above but using decreasing concentrations (125–15.625 μg/mL) of anthraquinone. The MIC value corresponds to the lowest concentration of the compound that completely inhibits the growth of the bacteria as detected by the unaided eye.

3.10. Evaluation of the Sensitizing Potential of DHDNA in Antibiotic-Resistant Bacteria

To gain preliminary insight into the potential of DHDNA to sensitize bacteria resistant to conventional antibiotics, the protocol described by Rangel et al. [70], with minor modifications, was used. One strain of each species (*S. aureus*, *E. faecalis*, and *E. coli*) was chosen from among the nine that were isolated and exposed to DHDNA dissolved in agar at the respective sub-MIC concentrations for each of these bacteria (15.625 μg/mL, 31.125 μg/mL, and 125 μg/mL). The bacteria selected were those that showed resistance to the greatest number of antibiotics in the sensitivity tests. Bacterial suspensions were prepared in saline to match the 0.5 McFarland turbidity standard and inoculated onto the agar surfaces using a sterile swab. Subsequently, antibiotic discs were incorporated, and the plates were incubated for 16 h at 37 °C. After incubation, the zones of inhibition around the discs on the control plates (with and without DMSO) were measured and compared with those on the plates with DHDNA.

3.11. Time-Kill Kinetic Assay

To identify whether the selected anthraquinone acts as a bacteriostatic or bactericidal agent, time-kill kinetic assays were performed. The procedure was as described by Huband et al. with minor modifications [71]. Experiments were performed using a log phase inoculum at a density of 1×10^6 cfu/mL cultured at 37 °C. Reference strains of *S. aureus*, as well as *E. faecalis*, were exposed to concentrations equal to their respective $1/2 \times$ MIC, $1 \times$ MIC, and $2 \times$ MIC. Aliquots of 100 μL were collected at 0, 1, 2, 4, 8, and 24 h of incubation and serially diluted in saline. Subsequently, 25 μL of each dilution was streaked on MHA plates and incubated for 16 h at 37 °C. To avoid the potential carry-over effect of the anthraquinone, the drops were allowed to dry before streaking them on the agar surfaces [72]. Finally, the number of colonies formed after incubation was counted on each plate. A compound is deemed bactericidal if it causes a reduction of $\geq$3 log10 in the number of colonies compared to the initial inoculum; conversely, when the reduction is $\leq$3 log10, the compound is classified as bacteriostatic [73].

All assays for antibacterial activity, MIC determination, sensitization potential assessment, and time-kill kinetics were performed at least three times as independent experiments.

3.12. Data Analysis and Visualizations

The 2D diagrams and 3D representations of the complexes were produced using the software Discovery Studio Visualizer-2021 and UCSF-Chimera-1.16. MD simulation analyses were visualized with GROMACS scripts in conjunction with Python scripts using the NumPy, Pandas, Matplotlib, Seaborn, and Pytraj libraries. The RMSD representations were generated from the alpha-carbon of the protein in the presence or absence of the ligands. Free energy binding calculations and time-kill kinetic assay were visualized using the NumPy, Pandas, and Matplotlib libraries.

4. Conclusions

In the search for new molecules showing antibacterial activity, naturally occurring anthraquinones have received more attention than synthetic derivatives. The in silico protocol we carried out to find unusual anthraquinones in nature with the potential to target the PPAT of *S. aureus*, *E. faecalis*, and *E. coli* identified 1,8-dihydroxy-4,5-dinitroanthraquinone (DHDNA) as a promising new inhibitor. In addition, these analyses suggested that nitro groups are critical, primarily for establishing interactions at the PPAT active site, increasing its affinity. In vitro experiments revealed that this compound exhibits a marked detrimental effect on the growth of both *S. aureus* and *E. faecalis* at lower concentrations than other reported anthraquinones and acts as a bacteriostatic agent. However, it was ineffective against *E. coli*, which could be associated with the known low permissiveness of Gram-negative bacteria to the entry of a wide range of antimicrobial agents. Furthermore, the lack of activity exhibited by 1,8-dihydroxyanthraquinone in both Gram-positive bacteria highlights the strong influence of nitro groups on the antibacterial effect of DHDNA. This is an important finding for the synthesis and evaluation of new nitro derivatives inspired by the anthraquinone structure. Collectively, the results of this work present DHDNA as a new antibacterial anthraquinone against *S. aureus* and *E. faecalis*, potentially acting as a PPAT inhibitor. Given that DHDNA demonstrated activity against antibiotic-resistant bacteria of both Gram-positive species and exhibited lower MIC values compared to several previously described anthraquinones, this compound deserves to be considered in future studies for further exploration of its antibacterial activity.

Supplementary Materials: The following supporting information can be downloaded at https://www.mdpi.com/article/10.3390/molecules29010203/s1: Supplementary Table S1. Virtual screening results of 114 anthraquinones on the nucleotide-binding site of PPAT of *S. aureus*, *E. faecalis*, and *E. coli*. Highlighted in different colors are the molecules commented on in the main text; Supplementary Table S2. Results of pharmacophore-based virtual screening of 114 anthraquinones ranked by RMSD values. Analysis performed with Pharmit; Supplementary Table S3. Binding free energy calculation of SaPPAT-DHDNA states extracted from the single trajectory of MD simulation. The calculations were conducted on the final 50 ns run using gmx-MMPBSA-1.5.7 by analyzing 500 snapshots; Supplementary Table S4. Binding free energy calculation of EfPPAT-DHDNA states extracted from the single trajectory of MD simulation. The calculations were conducted on the final 50 ns run using gmx-MMPBSA-1.5.7 by analyzing 500 snapshots; Supplementary Table S5. Binding free energy calculation of EcPPAT-DHDNA states extracted from the single trajectory of MD simulation. The calculations were conducted on the final 50 ns run using gmx-MMPBSA-1.5.7 by analyzing 500 snapshots; Supplementary Table S6. Pharmacokinetic profile of DHDNA. Analysis was performed with SwissADME; Supplementary Table S7. Target prediction of DHDNA using Swiss Target Prediction; Supplementary Table S8. Target prediction of Emodin using Swiss Target Prediction; Supplementary Table S9. Target prediction of Rhein using Swiss Target Prediction; Supplementary Table S10. Cytotoxic activity of DHDNA on NIH/3T3 cells (mouse embryonic fibroblast).

Author Contributions: Conceptualization, J.A. and J.C.; Methodology, J.A., V.V., A.C., M.M. and J.C.; Validation, J.A. and J.C.; Formal analysis, J.A., V.V., A.C., M.M. and J.C.; Investigation, J.A. and J.C.; Resources, J.A. and J.C.; Visualization, J.A.; Supervision, J.C.; Writing—original draft, J.A. and J.C.; Writing—review and editing, J.C.; Project administration, J.C.; Funding acquisition, J.A. and J.C. All authors have read and agreed to the published version of the manuscript.

Funding: This research was supported by the Universidad Católica de Cuenca with grant number PICVII19-58.

Institutional Review Board Statement: Not applicable.

Informed Consent Statement: Not applicable.

Data Availability Statement: All data generated or analyzed during this study are included in this published article and its Supplementary Information Files.

Conflicts of Interest: The authors declare no conflicts of interest.

References

1. Prestinaci, F.; Pezzotti, P.; Pantosti, A. Antimicrobial Resistance: A Global Multifaceted Phenomenon. *Pathog. Glob. Health* **2015**, *109*, 309–318. [CrossRef] [PubMed]
2. Hawkey, P.M.; Jones, A.M. The Changing Epidemiology of Resistance. *J. Antimicrob. Chemother.* **2009**, *64*, i3–i10. [CrossRef] [PubMed]
3. Ahmed, S.; Ahmed, M.Z.; Rafique, S.; Almasoudi, S.E.; Shah, M.; Jalil, N.A.C.; Ojha, S.C. Recent Approaches for Downplaying Antibiotic Resistance: Molecular Mechanisms. *Biomed. Res. Int.* **2023**, *2023*, 5250040. [CrossRef] [PubMed]
4. Vivas, R.; Barbosa, A.A.T.; Dolabela, S.S.; Jain, S. Multidrug-Resistant Bacteria and Alternative Methods to Control Them: An Overview. *Microb. Drug Resist.* **2019**, *25*, 890–908. [CrossRef] [PubMed]
5. Amorim, J.C.; Carpio, J.M. Alpha-Naphthoflavone as a Novel Scaffold for the Design of Potential Inhibitors of the APH(3′)-IIIa Nucleotide-Binding Site of Enterococcus Faecalis. *Microorganisms* **2023**, *11*, 2351. [CrossRef] [PubMed]
6. Lambert, P.A. Bacterial Resistance to Antibiotics: Modified Target Sites. *Adv. Drug Deliv. Rev.* **2005**, *57*, 1471–1485. [CrossRef] [PubMed]
7. Kumawat, M.; Nabi, B.; Daswani, M.; Viquar, I.; Pal, N.; Sharma, P.; Tiwari, S.; Devojit Kumar Sarma, S.S.; Kumar, M.; Kumawat, M.; et al. Role of Bacterial Efflux Pump Proteins in Antibiotic Resistance across Microbial Species. *Microb. Pathog.* **2023**, *181*, 106182. [CrossRef]
8. Ghai, I.; Ghai, S. Understanding Antibiotic Resistance via Outer Membrane Permeability. *Infect. Drug Resist.* **2018**, *11*, 523–530. [CrossRef]
9. Kim, D.W.; Thawng, C.N.; Choi, J.H.; Lee, K.; Cha, C.J. Polymorphism of Antibiotic-Inactivating Enzyme Driven by Ecology Expands the Environmental Resistome. *ISME J.* **2018**, *12*, 267–276. [CrossRef]
10. Miller, J.R.; Ohren, J.; Sarver, R.W.; Mueller, W.T.; De Dreu, P.; Case, H.; Thanabal, V. Phosphopantetheine Adenylyltransferase from Escherichia Coli: Investigation of the Kinetic Mechanism and Role in Regulation of Coenzyme A Biosynthesis. *J. Bacteriol.* **2007**, *189*, 8196–8205. [CrossRef]
11. Kim, K.H.; Lopez-Casillas, F.; Bai, D.H.; Luo, X.; Pape, M.E. Role of Reversible Phosphorylation of Acetyl-CoA Carboxylase in Long-Chain Fatty Acid Synthesis. *FASEB J.* **1989**, *3*, 2250–2256. [CrossRef] [PubMed]
12. Leonardi, R.; Zhang, Y.-M.; Charles, O.; Rock, S.J. Coenzyme A: Back in Action. *Prog. Lipid Res.* **2005**, *44*, 125–153. [CrossRef] [PubMed]
13. Gupta, A.; Sharma, P.; Singh, T.P.; Sharma, S. Phosphopantetheine Adenylyltransferase: A Promising Drug Target to Combat Antibiotic Resistance. *Biochim. Biophys. Acta–Proteins Proteom.* **2021**, *1869*, 140566. [CrossRef] [PubMed]
14. De Jonge, B.L.M.; Walkup, G.K.; Lahiri, S.D.; Huynh, H.; Neckermann, G.; Utley, L.; Nash, T.J.; Brock, J.; San Martin, M.; Kutschke, A.; et al. Discovery of Inhibitors of 4′-Phosphopantetheine Adenylyltransferase (PPAT) to Validate PPAT as a Target for Antibacterial Therapy. *Antimicrob. Agents Chemother.* **2013**, *57*, 6005–6015. [CrossRef] [PubMed]
15. Malik, E.M.; Christa, E. Müller Anthraquinones As Pharmacological Tools and Drugs. *Med. Res. Rev.* **2016**, *36*, 705–748. [CrossRef] [PubMed]
16. Malik, M.S.; Alsantali, R.I.; Jassas, R.S.; Alsimaree, A.A.; Syed, R.; Alsharif, M.A.; Kalpana, K.; Morad, M.; Althagafi, I.I.; Ahmed, S.A. Journey of Anthraquinones as Anticancer Agents-a Systematic Review of Recent Literature. *RSC Adv.* **2021**, *11*, 35806–35827. [CrossRef] [PubMed]
17. Li, Y.; Jiang, J.G. Health Functions and Structure-Activity Relationships of Natural Anthraquinones from Plants. *Food Funct.* **2018**, *9*, 6063–6080. [CrossRef]
18. Yusuf, M.; Mohammad, F.; Shabbir, M. Eco-Friendly and Effective Dyeing of Wool with Anthraquinone Colorants Extracted from Rubia Cordifolia Roots: Optimization, Colorimetric and Fastness Assay, Coloring Studies with Anthraquinone Colorants Extracted from Rubia Cordifolia Roots on Wool. *J. King Saud Univ.–Sci.* **2017**, *29*, 137–144. [CrossRef]
19. Zhang, M.-M.; Gong, Z.-C.; Zhao, Q.; Xu, D.-Q.; Fu, R.-J.; Tang, Y.-P.; Chen, Y.-Y. Time-Dependent Laxative Effect of Sennoside A, the Core Functional Component of Rhubarb, Is Attributed to Gut Microbiota and Aquaporins. *J. Ethnopharmacol.* **2023**, *311*, 116431. [CrossRef]
20. Sayed, H.M.; Ramadan, M.A.; Salem, H.H.; Ahmad, I.; Patel, H.; Fayed, M.A.A. Phytochemical Investigation, In Silico/In Vivo Analgesic, and Anti-Inflammatory Assessment of the Egyptian *Cassia occidentalis* L. *Steroids* **2023**, *196*, 109245. [CrossRef]
21. Kesharwani, D.; Das Paul, S.; Paliwal, R.; Satapathy, T. Exploring Potential of Diacerin Nanogel for Topical Application in Arthritis: Formulation Development, QbD Based Optimization and Pre-Clinical Evaluation. *Colloids Surf. B Biointerfaces* **2023**, *223*, 113160. [CrossRef] [PubMed]
22. Arrousse, N.; Harras, M.F.; El Kadiri, S.; Haldhar, R.; Ichou, H.; Bousta, D.; Grafov, A.; Rais, Z.; Taleb, M. New Anthraquinone Drugs and Their Anticancer Activities: Cytotoxicity, DFT, Docking and ADMET Properties. *Results Chem.* **2023**, *6*, 100996. [CrossRef]
23. Zhu, Y.; Yu, J.; Chen, T.; Liu, W.; Huang, Y.; Li, J.; Zhang, B.; Zhu, G.; He, Z.; Long, Y.; et al. Design, Synthesis, and Biological Evaluation of a Series of New Anthraquinone Derivatives as Anti-ZIKV Agents. *Eur. J. Med. Chem.* **2023**, *258*, 115620. [CrossRef] [PubMed]
24. Alias, C.; Feretti, D.; Viola, G.V.C.; Zerbini, I.; Bisceglie, F.; Pelosi, G.; Zani, C.; Buschini, A.; Carcelli, M.; Rogolino, D.; et al. Allium Cepa Tests: A Plant-Based Tool for the Early Evaluation of Toxicity and Genotoxicity of Newly Synthetized Antifungal Molecules. *Mutat. Res. Toxicol. Environ. Mutagen.* **2023**, *889*, 503654. [CrossRef]

25. Mahanty, S.; Rathinasamy, K. The Natural Anthraquinone Dye Purpurin Exerts Antibacterial Activity by Perturbing the FtsZ Assembly. *Bioorg. Med. Chem.* **2021**, *50*, 116463. [CrossRef] [PubMed]
26. Carpio Arévalo, J.M.; Amorim, J.C. An In-Silico Analysis Reveals 7,7′-Bializarin as a Promising DNA Gyrase B Inhibitor on Gram-Positive and Gram-Negative Bacteria. *Comput. Biol. Med.* **2021**, *135*, 104626. [CrossRef] [PubMed]
27. Amorim, J.C.; Cabrera Bermeo, A.E.; Vásquez, V.E.; Urgilés, M.R.M.; León, J.M.; Carpio, A. An Silico Evaluation of Anthraquinone Derivatives as Potential Inhibitors of DNA Gyrase B of Mycobacterium Tuberculosis. *Microorganisms* **2022**, *10*, 2434. [CrossRef]
28. Duan, F.; Xin, G.; Niu, H.; Huang, W. Chlorinated Emodin as a Natural Antibacterial Agent against Drug-Resistant Bacteria through Dual Influence on Bacterial Cell Membranes and DNA. *Sci. Rep.* **2017**, *7*, 12721. [CrossRef]
29. Wang, J.; Qu, Q.; Liu, X.; Cui, W.; Yu, F.; Chen, X.; Xing, X.; Zhou, Y.; Yang, Y.; Bello-Onaghise, G.; et al. 1-Hydroxyanthraquinone Exhibited Antibacterial Activity by Regulating Glutamine Synthetase of Staphylococcus Xylosus as a Virulence Factor. *Biomed. Pharmacother.* **2020**, *123*, 109779. [CrossRef]
30. Lu, C.; Wang, H.; Lv, W.; Xu, P.; Zhu, J.; Xie, J.; Liu, B.; Lou, Z. Antibacterial Properties of Anthraquinones Extracted from Rhubarb against Aeromonas Hydrophila. *Fish. Sci.* **2011**, *77*, 375–384. [CrossRef]
31. Boháčová, V.; Dočolomanský, P.; Breier, A.; Gemeiner, P.; Ziegelhöffer, A. Interaction of Lactate Dehydrogenase with Anthraquinone Dyes: Characterization of Ligands for Dye-Ligand Chromatography. *J. Chromatogr. B Biomed. Appl.* **1998**, *715*, 273–281. [CrossRef] [PubMed]
32. De Moliner, E.; Moro, S.; Sarno, S.; Zagotto, G.; Zanotti, G.; Pinna, L.A.; Battistutta, R. Inhibition of Protein Kinase CK2 by Anthraquinone-Related Compounds: A Structural Insight. *J. Biol. Chem.* **2003**, *278*, 1831–1836. [CrossRef] [PubMed]
33. Amaro, R.E.; Baudry, J.; Chodera, J.; Demir, Ö.; McCammon, J.A.; Miao, Y.; Smith, J.C. Ensemble Docking in Drug Discovery. *Biophys. J.* **2018**, *114*, 2271–2278. [CrossRef] [PubMed]
34. Seidel, T.; Ibis, G.; Bendix, F.; Wolber, G. Strategies for 3D Pharmacophore-Based Virtual Screening. *Drug Discov. Today Technol.* **2010**, *7*, e221–e228. [CrossRef] [PubMed]
35. Sunseri, J.; Koes, D.R. Pharmit: Interactive Exploration of Chemical Space. *Nucleic Acids Res.* **2016**, *44*, W442–W448. [CrossRef] [PubMed]
36. Szymańska, M.; Majerz, I. Effect of Substitution of Hydrogen Atoms in the Molecules of Anthrone and Anthraquinone. *Molecules* **2021**, *26*, 502. [CrossRef]
37. Chu, J.J.H.; Lee, R.C.H.; Ang, M.J.Y.; Wang, W.L.; Lim, H.A.; Wee, J.L.K.; Joy, J.; Hill, J.; Brian Chia, C.S. Antiviral Activities of 15 Dengue NS2B-NS3 Protease Inhibitors Using a Human Cell-Based Viral Quantification Assay. *Antivir. Res.* **2015**, *118*, 68–74. [CrossRef]
38. Song, Z.M.; Zhang, J.L.; Zhou, K.; Yue, L.M.; Zhang, Y.; Wang, C.Y.; Wang, K.L.; Xu, Y. Anthraquinones as Potential Antibiofilm Agents Against Methicillin-Resistant Staphylococcus Aureus. *Front. Microbiol.* **2021**, *12*, 709826. [CrossRef]
39. Celik, S.; Ozkok, F.; Ozel, A.E.; Müge Sahin, Y.; Akyuz, S.; Sigirci, B.D.; Kahraman, B.B.; Darici, H.; Karaoz, E. Synthesis, FT-IR and NMR Characterization, Antimicrobial Activity, Cytotoxicity and DNA Docking Analysis of a New Anthraquinone Derivate Compound. *J. Biomol. Struct. Dyn.* **2020**, *38*, 756–770. [CrossRef]
40. Friedman, M.; Xu, A.; Lee, R.; Nguyen, D.N.; Phan, T.A.; Hamada, S.M.; Panchel, R.; Tam, C.C.; Kim, J.H.; Cheng, L.W.; et al. The Inhibitory Activity of Anthraquinones against Pathogenic Protozoa, Bacteria, and Fungi and the Relationship to Structure. *Molecules* **2020**, *25*, 3101. [CrossRef]
41. Da Silva, R.A.G.; Wong, J.J.; Antypas, H.; Choo, P.Y.; Goh, K.; Jolly, S.; Liang, C.; Sing, L.T.K.; Veleba, M.; Hu, G.; et al. Mitoxantrone Targets Both Host and Bacteria to Overcome Vancomycin Resistance in Enterococcus Faecalis. *Sci. Adv.* **2023**, *9*, eadd9280. [CrossRef] [PubMed]
42. Manojlovic, N.T.; Novakovic, M.; Stevovic, V.; Solujic, S. Antimicrobial Metabolites from Three Serbian Caloplaca. *Pharm. Biol.* **2005**, *43*, 718–722. [CrossRef]
43. Kemegne, G.A.; Mkounga, P.; Essia Ngang, J.J.; Sado Kamdem, S.L.; Nkengfack, A.E. Antimicrobial Structure-Activity Relationship of Five Anthraquinones of Emodine Type Isolated from Vismia Laurentii. *BMC Microbiol.* **2017**, *17*, 41. [CrossRef] [PubMed]
44. Peerzada, Z.; Kanhed, A.M.; Desai, K.B. Effects of Active Compounds from Cassia Fistula on Quorum Sensing Mediated Virulence and Biofilm Formation in Pseudomonas Aeruginosa. *RSC Adv.* **2022**, *12*, 15196–15214. [CrossRef] [PubMed]
45. Breijyeh, Z.; Jubeh, B.; Karaman, R. Resistance of Gram-Positive Bacteria to Current Antibacterial Agents and Overcoming Approaches. *Molecules* **2020**, *25*, 2888. [CrossRef] [PubMed]
46. Liu, M.; Peng, W.; Qin, R.; Yan, Z.; Cen, Y.; Zheng, X.; Pan, X.; Jiang, W.; Li, B.; Li, X.; et al. The Direct Anti-MRSA Effect of Emodin via Damaging Cell Membrane. *Appl. Microbiol. Biotechnol.* **2015**, *99*, 7699–7709. [CrossRef] [PubMed]
47. Alves, D.S.; Pérez-Fons, L.; Estepa, A.; Micol, V. Membrane-Related Effects Underlying the Biological Activity of the Anthraquinones Emodin and Barbaloin. *Biochem. Pharmacol.* **2004**, *68*, 549–561. [CrossRef]
48. Richter, M.F.; Drown, B.S.; Riley, A.P.; Garcia, A.; Shirai, T.; Svec, R.L.; Hergenrother, P.J. Predictive Compound Accumulation Rules Yield a Broad-Spectrum Antibiotic. *Nature* **2017**, *545*, 299–304. [CrossRef]
49. Richter, M.F.; Hergenrother, P.J. The Challenge of Converting Gram-Positive-Only Compounds into Broad-Spectrum Antibiotics. *Ann. N. Y. Acad. Sci.* **2019**, *1435*, 18–38. [CrossRef]

50. Azelmat, J.; Larente, J.F.; Grenier, D. The Anthraquinone Rhein Exhibits Synergistic Antibacterial Activity in Association with Metronidazole or Natural Compounds and Attenuates Virulence Gene Expression in Porphyromonas Gingivalis. *Arch. Oral Biol.* **2015**, *60*, 342–346. [CrossRef]
51. Machado, F.P.; Rodrigues, I.C.; Gales, L.; Pereira, J.A.; Costa, P.M.; Dethoup, T.; Mistry, S.; Silva, A.M.S.; Vasconcelos, V.; Kijjoa, A. New Alkylpyridinium Anthraquinone, Isocoumarin, C-Glucosyl Resorcinol Derivative and Prenylated Pyranoxanthones from the Culture of a Marine Sponge-Associated Fungus, Aspergillus Stellatus KUFA 2017. *Mar. Drugs* **2022**, *20*, 672. [CrossRef] [PubMed]
52. Dell'Annunziata, F.; Folliero, V.; Palma, F.; Crudele, V.; Finamore, E.; Sanna, G.; Manzin, A.; De Filippis, A.; Galdiero, M.; Franci, G. Anthraquinone Rhein Exhibits Antibacterial Activity against Staphylococcus Aureus. *Appl. Sci.* **2022**, *12*, 8691. [CrossRef]
53. Irwin, J.J.; Tang, K.G.; Young, J.; Dandarchuluun, C.; Wong, B.R.; Khurelbaatar, M.; Moroz, Y.S.; Mayfield, J.; Sayle, R.A. ZINC20—A Free Ultralarge-Scale Chemical Database for Ligand Discovery. *J. Chem. Inf. Model.* **2020**, *60*, 6065–6073. [CrossRef] [PubMed]
54. Hanwell, M.D.; Curtis, D.E.; Lonie, D.C.; Vandermeersch, T.; Eva Zurek, G.R.H. Avogadro: An Advanced Semantic Chemical Editor, Visualization, and Analysis Platform. *J. Cheminform.* **2012**, *4*, 17. [CrossRef] [PubMed]
55. Sander, T.; Freyss, J.; Von Korff, M.; Rufener, C. DataWarrior: An Open-Source Program for Chemistry Aware Data Visualization and Analysis. *J. Chem. Inf. Model.* **2015**, *55*, 460–473. [CrossRef] [PubMed]
56. Yoon, H.J.; Kang, J.Y.; Mikami, B.; Lee, H.H.; Suh, S.W. Crystal Structure of Phosphopantetheine Adenylyltransferase from Enterococcus Faecalis in the Ligand-Unbound State and in Complex with ATP and Pantetheine. *Mol. Cells* **2011**, *32*, 431–435. [CrossRef] [PubMed]
57. Moreau, R.J.; Skepper, C.K.; Appleton, B.A.; Blechschmidt, A.; Balibar, C.J.; Benton, B.M.; Drumm, J.E., III; Feng, B.Y.; Geng, M.; Li, C.; et al. Fragment-Based Drug Discovery of Inhibitors of Phosphopantetheine Adenylyltransferase from Gram-Negative Bacteria. *J. Med. Chem.* **2018**, *61*, 3309–3324. [CrossRef]
58. Pettersen, E.F.; Goddard, T.D.; Huang, C.C.; Couch, G.S.; Greenblatt, D.M.; Meng, E.C.; Ferrin, T.E. UCSF Chimera—A Visualization System for Exploratory Research and Analysis. *J. Comput. Chem.* **2004**, *25*, 1605–1612. [CrossRef]
59. Korb, O.; Stützle, T.; Exner, T.E. PLANTS: Application of Ant Colony Optimization to Structure-Based Drug Design. In *International Workshop on Ant Colony Optimization and Swarm Intelligence*; Springer: Berlin/Heidelberg, Germany, 2006; Volume 4150, pp. 247–258. [CrossRef]
60. Van Der Spoel, D.; Lindahl, E.; Hess, B.; Groenhof, G.; Mark, A.E.; Berendsen, H.J.C. GROMACS: Fast, Flexible, and Free. *J. Comput. Chem.* **2005**, *26*, 1701–1718. [CrossRef]
61. Best, R.B.; Zhu, X.; Shim, J.; Lopes, P.E.M.; Mittal, J.; Feig, M.; MacKerell, A.D. Optimization of the Additive CHARMM All-Atom Protein Force Field Targeting Improved Sampling of the Backbone φ, ψ and Side-Chain X1 and X2 Dihedral Angles. *J. Chem. Theory Comput.* **2012**, *8*, 3257–3273. [CrossRef]
62. Berendsen, H.J.C.; Postma, J.P.M.; Van Gunsteren, W.F.; Dinola, A.; Haak, J.R. Molecular Dynamics with Coupling to an External Bath. *J. Chem. Phys.* **1984**, *81*, 3684–3690. [CrossRef]
63. Ewald, P.P. Die Berechnung Optischer Und Elektrostatischer Gitterpotentiale. *Ann. Phys.* **1921**, *369*, 253–287. [CrossRef]
64. Hess, B.; Bekker, H.; Berendsen, H.J.C.; Fraaije, J.G.E.M. LINCS: A Linear Constraint Solver for Molecular Simulations. *J. Comput. Chem.* **1997**, *18*, 1463–1472. [CrossRef]
65. Kollman, P.A.; Massova, I.; Reyes, C.; Kuhn, B.; Huo, S.; Chong, L.; Lee, M.; Lee, T.; Duan, Y.; Wang, W.; et al. Calculating Structures and Free Energies of Complex Molecules: Combining Molecular Mechanics and Continuum Models. *Acc. Chem. Res.* **2000**, *33*, 889–897. [CrossRef] [PubMed]
66. Valdés-Tresanco, M.S.; Valdés-Tresanco, M.E.; Valiente, P.A.; Moreno, E. Gmx_MMPBSA: A New Tool to Perform End-State Free Energy Calculations with GROMACS. *J. Chem. Theory Comput.* **2021**, *17*, 6281–6291. [CrossRef]
67. Daina, A.; Michielin, O.; Zoete, V. SwissADME: A Free Web Tool to Evaluate Pharmacokinetics, Drug-Likeness and Medicinal Chemistry Friendliness of Small Molecules. *Sci. Rep.* **2017**, *7*, 42717. [CrossRef] [PubMed]
68. Daina, A.; Michielin, O.; Zoete, V. SwissTargetPrediction: Updated Data and New Features for Efficient Prediction of Protein Targets of Small Molecules. *Nucleic Acids Res.* **2019**, *47*, W357–W3664. [CrossRef]
69. Varsou, D.D.; Melagraki, G.; Sarimveis, H.; Afantitis, A. MouseTox: An Online Toxicity Assessment Tool for Small Molecules through Enalos Cloud Platform. *Food Chem. Toxicol.* **2017**, *110*, 83–93. [CrossRef]
70. Azucena, R.C.I.; Roberto, C.L.J.; Martin, Z.R.; Rafael, C.Z.; Leonardo, H.H.; Gabriela, T.P.; Araceli, C.R. Drug Susceptibility Testing and Synergistic Antibacterial Activity of Curcumin with Antibiotics against Enterotoxigenic Escherichia Coli. *Antibiotics* **2019**, *8*, 43. [CrossRef]
71. Huband, M.D.; Bradford, P.A.; Otterson, L.G.; Basarab, G.S.; Kutschke, A.C.; Giacobbe, R.A.; Patey, S.A.; Alm, R.A.; Johnstone, M.R.; Potter, M.E.; et al. In Vitro Antibacterial Activity of AZD0914, a New Spiropyrimidinetrione DNA Gyrase/Topoisomerase Inhibitor with Potent Activity against Gram-Positive, Fastidious Gram-Negative, and Atypical Bacteria. *Antimicrob. Agents Chemother.* **2015**, *59*, 467–474. [CrossRef]

72. Garrigós, C.; Murillo, O.; Lora-Tamayo, J.; Verdaguer, R.; Tubau, F.; Cabellos, C.; Cabo, J.; Ariza, J. Fosfomycin-Daptomycin and Other Fosfomycin Combinations as Alternative Therapies in Experimental Foreign-Body Infection by Methicillin-Resistant Staphylococcus Aureus. *Antimicrob. Agents Chemother.* **2013**, *57*, 606–610. [CrossRef] [PubMed]
73. Belley, A.; Neesham-Grenon, E.; Arhin, F.F.; McKay, G.A.; Parr, T.R.; Moeck, G. Assessment by Time-Kill Methodology of the Synergistic Effects of Oritavancin in Combination with Other Antimicrobial Agents against Staphylococcus Aureus. *Antimicrob. Agents Chemother.* **2008**, *52*, 3820–3822. [CrossRef] [PubMed]

molecules

Review

Machine Learning Empowering Drug Discovery: Applications, Opportunities and Challenges

Xin Qi [1,*,†], Yuanchun Zhao [1,†], Zhuang Qi [2], Siyu Hou [1] and Jiajia Chen [1]

[1] School of Chemistry and Life Sciences, Suzhou University of Science and Technology, Suzhou 215011, China; 2111121006@post.usts.edu.cn (Y.Z.); 2211121003@post.usts.edu.cn (S.H.); njucjj@126.com (J.C.)
[2] School of Software, Shandong University, Jinan 250101, China; z_qi@mail.sdu.edu.cn
* Correspondence: qixin@usts.edu.cn; Tel.: +86-0512-68418434
† These authors contributed equally to this work.

Abstract: Drug discovery plays a critical role in advancing human health by developing new medications and treatments to combat diseases. How to accelerate the pace and reduce the costs of new drug discovery has long been a key concern for the pharmaceutical industry. Fortunately, by leveraging advanced algorithms, computational power and biological big data, artificial intelligence (AI) technology, especially machine learning (ML), holds the promise of making the hunt for new drugs more efficient. Recently, the Transformer-based models that have achieved revolutionary breakthroughs in natural language processing have sparked a new era of their applications in drug discovery. Herein, we introduce the latest applications of ML in drug discovery, highlight the potential of advanced Transformer-based ML models, and discuss the future prospects and challenges in the field.

Keywords: machine learning; drug discovery; transformer; opportunity; challenge

1. Introduction

Drug research and development play a vital role in improving human health and well-being. However, the discovery of a new drug is an extremely complex, expensive and time-consuming process, typically costing approximately USD 2.6 billion [1] and taking more than 10 years on average [2]. Despite the high investment levels, the approval success rate of launching a small-molecule drug to market from phase I clinical trial is less than 10% [3], highlighting the considerable risk of failure. Therefore, how to reduce the costs and accelerate the pace of new drug discovery has emerged as a key concern within the pharmaceutical industry.

The increasing availability of large-scale biomedical data provides tremendous opportunities for computational drug discovery, but effectively mining, correlating, and analyzing these huge amounts of data has become a critical challenge. Fortunately, with the advent of efficient mathematical tools and abundant computational resources, artificial intelligence (AI) approaches have rapidly developed (Figure 1). As the representative AI method, machine learning (ML), empowers machines to learn from existing data by using statistical approaches and make predictions, which can be further classified into supervised, unsupervised, and reinforcement learnings [4,5]. Deep learning (DL), a subset of ML, focuses on using multi-layered artificial neural networks (ANNs) structures to simulate the neural networks of the human brain for learning data representations, making it more powerful and flexible in handling complex and high-dimensional data [6,7]. With the advantages of low cost and fast speed, the ML approaches are revolutionizing and strengthening multiple stages of drug discovery, such as target identification, *de novo* drug design and drug repurposing. For example, DL-based open-source tools, such as DeepDTAF [8] and DeepAffinity [9], have been applied to predict the binding affinity of drug–target interactions (DTIs), making the hunt for new pharmaceuticals more efficient. Accordingly, more and more pharmaceutical giants, such as Sanofi (Paris, France), Merck (Darmstadt,

Germany), Takeda (Takeda, Japan) and Genentech (South San Francisco, America), have initiated cooperation with AI companies to develop new drugs.

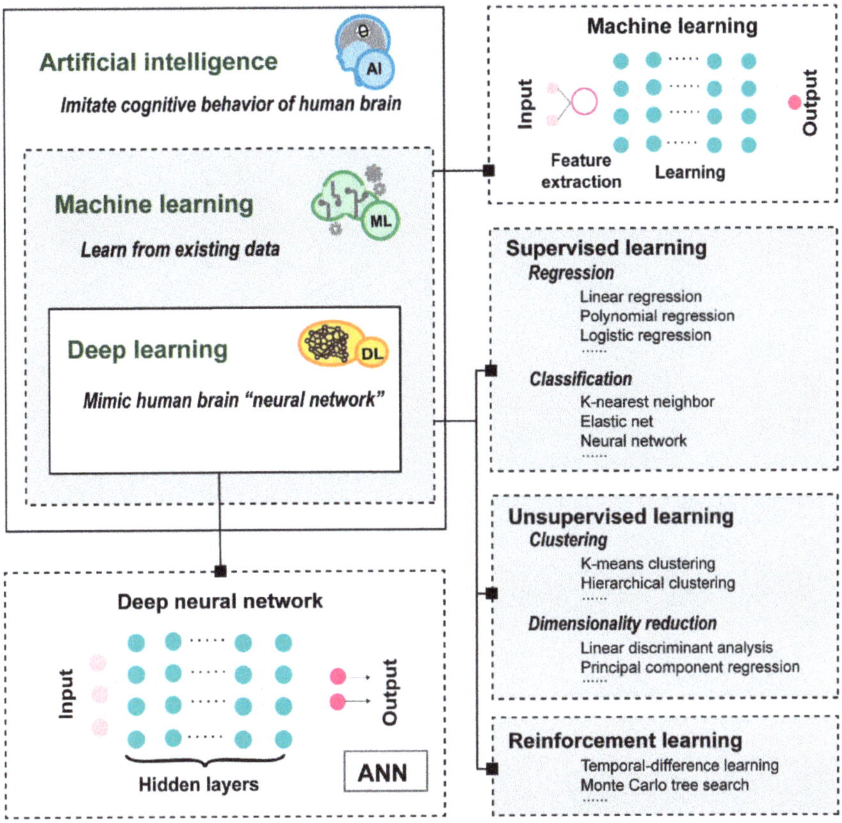

Figure 1. Introduction diagram of artificial intelligence and its subfields: machine learning and deep learning.

Notably, the Transformer-based language models, such as the Generative Pre-training Transformer (GPT), Bidirectional Encoder Representations from Transformers (BERT) and the Text-to-Text Transfer Transformer (T5), have not only achieved revolutionary breakthroughs but have also brought about a paradigm shift in the area of natural language processing (NLP) [10]. In particular, the outstanding learning ability, generalization ability and transferability of Transformer-based language models have sparked a new era of their applications in drug discovery and development, primarily owing to the inherent similarities between drug-related biological sequences and natural languages. Their remarkable advantages, including capturing long-range dependencies in sequences, processing input sequences in parallel, employing an attention mechanism, and having extendibility to incorporate multimodal information, make them valuable tools for various aspects of the drug discovery process [11]. For example, by employing Transformer-based language models, Kalakoti et al. [12] have successfully developed a modular framework called TransDTI for predicting novel DTIs from sequence data. Its performance proved to be superior to existing methods. Therefore, the Transformer-based models have the potential to revolutionize the identification and development of new drugs.

Given the significance of ML techniques in the pharmaceutical industry, we here focus on introducing the recent advancements, opportunities and challenges of ML applications in drug discovery. First, we provide an updated overview of the emerging applications of ML in different stages of the drug discovery process, including drug design, drug screening, drug repurposing and chemical synthesis. Next, we highlight the opportunities of the advanced Transformer-based models in empowering drug discovery. Furthermore, we discuss the challenges and future prospects of ML in the field of drug discovery.

2. Applications of ML in Drug Discovery

The process of discovering effective new drugs is time-consuming and predominantly the most challenging part of drug development. With the advantages in learning from data, discerning patterns, and making intelligent decisions, ML-based approaches have emerged as versatile tools that can be applied in multiple stages of drug discovery, including drug design, drug screening, drug repurposing and chemical synthesis (Figure 2). Moreover, considerable efforts are dedicated to developing models, tools, software and databases based on the core architecture of ML algorithms, to counter the inefficiencies and uncertainties inherent in traditional drug development methods (Table 1).

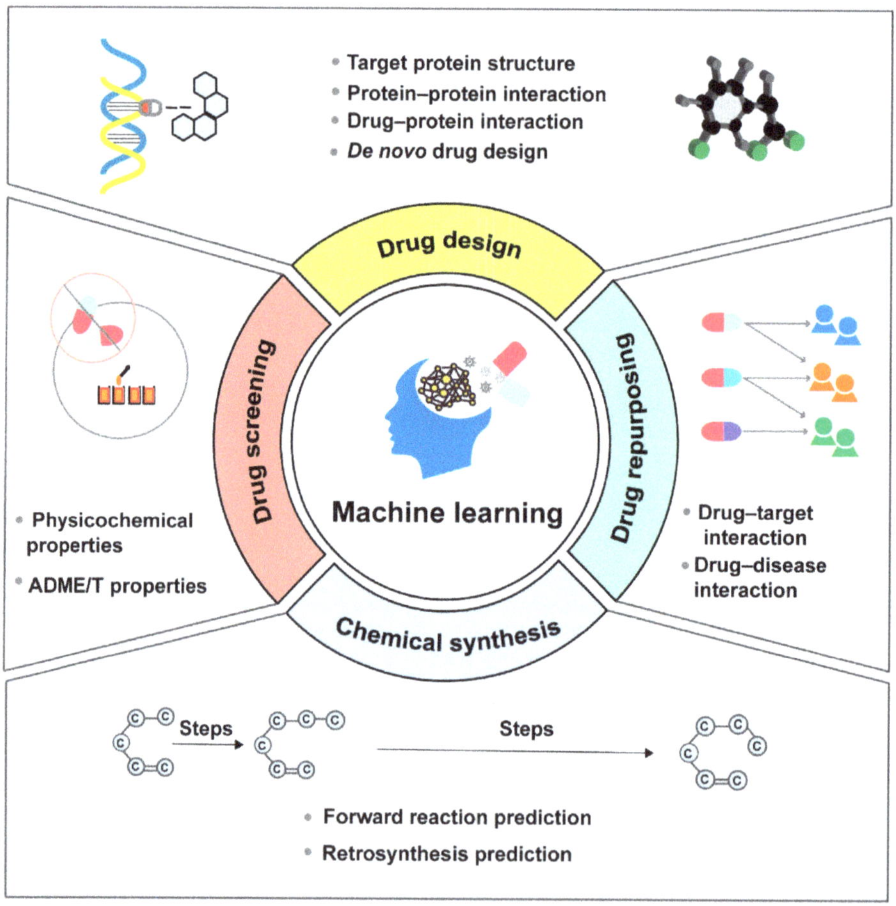

Figure 2. Machine learning can be applied in multiple stages of the drug discovery process, mainly including drug design, drug screening, drug repurposing and chemical synthesis.

Table 1. ML-based software/model used for drug discovery.

Name	Algorithm	Specific Function	PMID
Prediction of the target protein structure			
TrRosetta server	DNN	Predict 3D structures of proteins	[13]
AlphaFold	DNN	Predict 3D structures of proteins	[14]
ComplexQA	GNN	Predict protein complex structure	[15]
ProteinBERT	Transformer	Predict secondary structure	[16]
ESMfold	Transformer	Predict structure of proteins	[17]
Predicting protein–protein interactions			
IntPred	RF	Predict PPI interface sites	[18]
eFindSite	SVM; NBC	Predict PPI interfaces	[19]
DELPHI	RNN; CNN	Predict PPI sites	[20]
PPISP-XGBoost	XGBoost	Predict PPI sites	[21]
HN-PPISP	CNN	Predict PPI sites	[22]
TAGPPI	GCN	Predict PPIs	[23]
Struct2Graph	GAT	Predict PPIs	[24]
DeepFE-PPI	DNN	Predict PPIs	[25]
SGPPI	GCN	Predict PPIs	[26]
DeepPPI	DNN	Predict PPIs	[27]
DL-PPI	GNN	Predict PPIs	[28]
DeepSG2PPI	CNN	Predict PPIs	[29]
MaTPIP	Transformer; CNN	Predict PPIs	[30]
ProtInteract	Autoencoder; CNN	Predict PPIs	[31]
Predicting drug–target interactions			
DeepC-SeqSite	CNN	Predict DTI binding sites	[32]
DeepSurf	CNN; ResNet	Predict DTI binding sites	[33]
PrankWeb	RF	Predict DTI binding sites	[34]
PUResNet	ResNet	Predict DTI binding sites	[35]
AGAT-PPIS	GNN	Predict DTI binding sites	[36]
DeepDTA	CNN	Predict DTI binding affinity	[37]
SimBoost	GBM	Predict DTI binding affinity	[38]
DEELIG	CNN	Predict DTI binding affinity	[39]
DeepDTAF	CNN	Predict DTI binding affinity	[8]
GraphDelta	CNN	Predict DTI binding affinity	[40]
PotentialNet	CNN	Predict DTI binding affinity	[41]
DeepAffinity	RNN, CNN	Predict DTI binding affinity	[9]
TeM-DTBA	CNN	Predict DTI binding affinity	[42]
Wang et al.'s method	RL	Predict DTI binding pose	[43]
Nguyen et al.'s method	RF; CNN	Predict DTI binding pose	[44]
AMMVF-DTI	GAT; NTN	Predict drug–target interactions	[45]
***De novo* drug design**			
ReLeaSE	RNN; RL	Conduct *de novo* drug design	[46]
ChemVAE	CNN; GRU	Conduct de novo drug design	[47]
MolRNN	RNN	Conduct multi-objective *de novo* drug design	[48]
PaccMann(RL)	VAE	Generate compounds with anti-cancer drug properties	[49]
druGAN	AAE	Conduct *de novo* drug design	[50]
SCScore	CNN	Evaluate the molecular accessibility	[51]
UnCorrupt SMILES	Transformer	Conduct *de novo* drug design	[52]
PETrans	Transfer learning	Conduct *de novo* drug design	[53]
FSM-DDTR	Transformer	Conduct *de novo* drug design	[54]
DNMG	GAN	Conduct *de novo* drug design	[55]
MedGAN	GAN	Design novel molecule	[56]
Prediction of the physicochemical properties			
Panapitiya et al.'s method	GNN	Predict aqueous solubility	[57]
SolTranNet	Transformer	Predict aqueous solubility	[58]
Zang et al.'s method	SVM	Predict multiple physicochemical properties	[59]
Prediction of the ADME/T properties			
ADMETboost	XGBoost	Predict ADME/T properties	[60]
vNN	k-NN	Predict ADME/T properties	[61]
Interpretable-ADMET	CNN; GAT	Predict ADME/T properties	[62]
XGraphBoost	GNN	Predict ADME/T properties	[63]
DeepTox	DNN	Predict toxicity of compounds	[64]
Li et al.'s method	DNN	Predict human Cytochrome P450 inhibition	[65]
LightBBB	LightGBM	Predict blood–brain barrier	[66]
Deep-B3	CNN	Predict blood–brain barrier	[67]
PredPS	GNN	Predict stability of compounds in human plasma	[68]
Khaouane et al.'s method	CNN	Predict plasma protein binding	[69]
Application of AI in drug repurposing			
deepDTnet	Autoencoder	Predict new targets of known drugs	[70]
NeoDTI	GCN	Predict new targets of known drugs	[71]
DTINet	Network diffusion algorithm and the dimensionality reduction	Predict new targets of known drugs	[72]
MBiRW	Birandom walk algorithm	Predict new indications of known drugs	[73]
GDRnet	GNN	Predict new indications of known drugs	[74]
deepDR	VAE	Predict new indications of known drugs	[75]
GIPAE	VAE	Predict new indications of known drugs	[76]
DrugRep-HeSiaGraph	Heterogeneous siamese neural network	Predict new indications of known drugs	[77]
iEdgeDTA	GCNN	Predict DTI binding affinity	[78]

Table 1. *Cont.*

Name	Algorithm	Specific Function	PMID
Retrosynthesis prediction			
Segler et al.'s method	MCTS, DNN	Predict retrosynthetic analysis	[79]
Liu et al.'s method	RNN	Predict retrosynthetic analysis	[80]
RAscore	RF	Predict retrosynthetic accessibility score	[81]
Reaction prediction			
Wei et al.'s method	Neural network	Predict reaction classes	[82]
Coley et al.'s method	Neural network	Predict products of chemical reactions	[83]
Gao et al.'s method	Neural network	Predict optimal reaction conditions	[84]
Marcou et al.'s method	RF	Evaluate reaction feasibility	[85]

Note: DNN, deep neural network; RNN, recurrent neural network; RF, random forest; CNN, convolutional neural network; XGBoost, eXtreme gradient boosting; GCN, graph convolutional network; GAT, graph attention network; SVM, support vector machine; NBC, naïve Bayes classifier; ResNet, residual network; GBM, gradient boosting machines; RL, reinforcement learning; GRU, gated recurrent unit; VAE, variational autoencoder; AAE, adaptive adversarial autoencoder; GNN, graph neural networks; k-NN, k-nearest neighbor; LightGBM, light gradient boosting machine; MCTS, Monte Carlo tree search, NTN, neural tensor network; GAN, generative adversarial network; GCNN, graph convolutional neural network.

2.1. Application of ML in Drug Design

2.1.1. Prediction of the Target Protein Structure

Since proteins play crucial roles in various biological processes, their dysfunctions can lead to abnormal cell behavior and lead to the development of diseases [86]. For selective targeting of diseases, small-molecule compounds are generally designed based on the three-dimensional (3D) chemical environment surrounding the ligand-binding sites of the target protein [87]. Hence, predicting the 3D structure of the target protein is of great significance for structure-based drug discovery. Homology modeling has traditionally been used for this purpose, relying on known protein structures as templates [88]. Comparatively, ML-based approaches have shown great promise in predicting the 3D structures of target proteins with improved accuracy and efficiency. For example, AlphaFold is a state-of-the-art protein structure prediction system developed by DeepMind, a leading AI company. Based on deep neural network (DNN), it has achieved remarkable success in multiple protein structure prediction competitions, demonstrating its ability to accurately predict the 3D structures of proteins by analyzing the adjacent amino acid distances and peptide bond angles [14]. Importantly, AlphaFold has significantly advanced the field of protein structure prediction and has the potential to revolutionize drug discovery [14]. Therefore, ML-based approaches hold great potential to enhance our understanding of protein structures. It should be noted that protein structures can undergo changes in different environments, and proteins may form multiple coexisting structures under the same conditions [89]. This complexity adds to the challenges of structure prediction.

2.1.2. Prediction of PPIs

In most cases, proteins rarely implement their functions alone, but rather cooperate with other proteins to form intricate relationships known as the protein–protein interaction (PPI) network [86]. PPIs possess indispensable functions in diverse biological processes. They can contribute to altering protein specificity, regulating protein activity and generating novel binding sites for effector molecules [90]. Hence, understanding and targeting PPIs offers opportunities to design innovative drugs that can modulate complex biological processes.

Currently, ML-based methods for PPI prediction can be broadly grouped into structure-based and sequence-based categories. Structure-based approaches mainly leverage the knowledge of protein structure similarity to predict PPIs [91]. For example, IntPred, a random forest ML tool, was developed to predict protein–protein interface sites based on structural features. Compared with other methods, the IntPred predictor showed strong performance in identifying interactions at both the surface-patch and residue levels on independent test sets of both obligate and transient complexes (Matthews' Correlation Coefficient (MCC) = 0.370, accuracy = 0.811, specificity = 0.916, sensitivity = 0.411) [18]. Struct2Graph, a graph attention network (GAT)-based classifier, was proposed to identify PPIs directly from the 3D structure of protein chains [24]. The accuracy of Struct2Graph on balanced sets with equal numbers of positive and negative pairs was 0.9989, and the average accuracy of five-fold cross-validation on unbalanced sets with a ratio of positive and negative pairs of 1:10 was 0.9942 [24]. Comparatively, sequence-based PPI prediction approaches aim to identify physical interactions between two proteins by leveraging information from their protein sequences [92]. DNNs provide a robust solution for this purpose. They are composed of multiple layers of interconnected neurons, allowing them to automatically extract complex patterns and features from data. For example, DeepPPI applied DNNs to effectively learn protein representations from common protein descriptors, thereby contributing to the prediction of PPIs. It can achieve excellent performance on the *S. cerevisiae* dataset with an accuracy of 0.925, precision of 0.9438, recall of 0.9056, specificity of 0.9449, MCC of 0.8508 and area under the curve (AUC) of 0.9743, respectively [27]. Extensive experiments showed that DeepPPI was able to learn the useful features of protein pairs through a layer-wise abstraction, resulting in better predictive performance than existing methods on core *S. cerevisiae*, *H. pylori* and *H. sapien* datasets [27]. In addition, based on Uniprot database, Li et al. [20] developed a DELPHI, a new sequence-based deep ensemble model for PPI-binding sites' prediction. Therefore, ML-based approaches have great potential in enhancing the identification of PPI sites. Compared with sequence-based approaches, structure-based ones are limited by the scarcity of available protein structures and the low quality of familiar protein structures [90,93].

2.1.3. Prediction of DTIs

Most drugs exert therapeutic effects by specifically interacting with target molecules within the body, such as enzymes, receptors and ion channels. Hence, the accurate prediction of DTIs is a pivotal step in the drug design pipeline. As the traditional experimental approaches are time-consuming and costly, ML-based methods have been increasingly developed and applied by researchers to predict DTIs. These methods primarily focus on three key aspects: predicting the binding sites of drugs on target molecules, estimating the binding affinity between drugs and targets, and determining the binding pose or conformation of the drug within the target molecule [94].

Firstly, binding sites, also referred to as binding pockets, are specific locations within a protein where interactions occur between the protein and a ligand (such as a drug molecule) [94]. By introducing a deep convolutional neural network (CNN), Cui et al. [32] developed a sequence-based method, DeepC-SeqSite, for predicting protein–ligand binding residues. Notably, this method exhibited superior performance compared with multiple existing sequence-based and 3D-structure-based methods, including the leading ligand-binding method COACH [32]. Similarly, Zhou et al. [36] proposed a binding site prediction method called AGAT-PPIS based on augmented GAT. It demonstrated significant improvements over the state-of-the-art method, achieving an accuracy increase of 8% on the benchmark test set. Moreover, binding affinity represents the strength of an interaction between a drug and its target. Some tools based on ML and DL algorithms have been applied to determine DTIs' binding affinity, such as DEELIG [39] and GraphDelta [40]. In addition, the active conformation of ligands plays a crucial role in facilitating the effective binding between proteins and drugs [94]. By combining random forest and CNN strategies, Nguyen et al. [44] proposed a scoring function to select the most relevant poses generated by docking software tools including GOLD, GLIDE and Autodock Vina, thereby contributing to obtaining more accurate and effective ligand–target binding configurations. Therefore, ML algorithms have been extensively employed to predict DTIs and hold the potential to facilitate the design of new drugs.

2.1.4. *De Novo* Drug Design

De novo drug design refers to the process of creating new drug molecules from scratch using computational methods, without relying on existing bioactive compounds or known drug structures. It involves designing molecules that have specific properties and functions to target a particular disease or condition [95,96]. Compounds developed with traditional *de novo* drug design methods (e.g., the fragment-based approach) usually have poor drug metabolism and pharmacokinetics properties and are hard to synthesize due to the complexity and impracticality of compound structures [97,98]. Therefore, there is high demand for new methods to explore chemical entities that meet the requirements of biological activity, drug metabolism, pharmacokinetics and synthesis practicality.

Recently, ML-based approaches, especially auto-encoder variants (e.g., the variational auto-encoder (VAE) and adversarial auto-encoder (AAE)) have gained attention in the field of *de novo* drug design. PaccMann^{RL} is an example of these approaches that combines a hybrid VAE with reinforcement learning for the *de novo* design of anti-cancer molecules from transcriptomic data [49]. Similarly, another approach, known as druGAN, utilizes a deep generative AAE model to generate novel molecules that possess specific anti-cancer properties [50]. In addition, a Wasserstein GAN and GCN-based model, known as MedGAN, has been successfully developed to generate novel quinoline-scaffold molecules from complicated molecular graphs and evaluate drug-related properties [56]. It has been demonstrated that the MedGAN was able to produce 25% effective molecules, 62% fully connected, among which 92% are quinoline, 93% are novel, and 95% are unique [56]. To address the difficulty in synthesizing generated molecules, Coley et al. [51] defined a synthetic complexity score, namely SCScore, that utilizes precedent reaction knowledge to train a neural network model for evaluating the level of synthetic complexity. Therefore, ML-empowering approaches play crucial roles in *de novo* drug design, revolutionizing the process of discovering and developing new drugs.

2.2. *Application of ML in Drug Screening*

2.2.1. Prediction of the Physicochemical Properties

The physicochemical properties of drugs, mainly including solubility, ionization degree, partition coefficient, permeability coefficient and stability, play a significant role in determining their behavior (e.g., bioavailability, absorption, transportation and permeability) in biological systems as well as the environment, and in evaluating their potential risks to human health [6,59]. Hence, these properties are assessed during drug screening to select promising candidates for further development and optimization. At present, multiple ML-based tools have been proposed to predict the physicochemical properties of molecules. For example, Francoeur et al. [58] developed a molecule attention Transformer called SolTranNet for predicting aqueous solubility from the SMILES representation of drug molecules. It has been demonstrated to function as a classifier for filtering insoluble compounds, achieving a sensitivity of 0.948 on Challenge to Predict Aqueous Solubility (SC2) datasets, which is competitive with other methods [58]. Moreover, by using molecular fingerprints and four ML algorithms, Zang et al. [59] developed a quantitative structure–property relationship workflow to predict six physicochemical properties of environmental chemicals, such as water solubility, octanol–water partition coefficient, melting point, boiling point, bioconcentration factor, and vapor pressure [59]. Therefore, these ML-based predictors are valuable tools in drug discovery, as they can help in screening potential drug candidates based on their physicochemical properties.

2.2.2. Prediction of the ADME/T Properties

Once hit or lead compounds are identified during the drug discovery process, a series of tests and evaluations are conducted to assess their absorption, distribution, metabolism, and excretion and toxicity (ADME/T) properties [99]. These pharmacokinetic properties are essential for understanding how the compounds will behave in the human body and whether they have the potential to be safe and effective as drugs. Imbalanced ADME/T properties frequently cause the failure of drug candidates in late stages of drug development and may even lead to the withdrawal of approved drugs [100]. Hence, ADME/T properties are often employed as molecular filters to screen large databases of compounds in the early stage of drug discovery, thereby helping to increase efficiency and improve the success rate of drug screening [93,100].

To detect the ADME/T properties of drugs, various evaluation criteria such as hepatotoxicity, passing through the blood–brain barrier (BBB), plasma protein binding (PPB) and cytochrome P450 2D6 (CYP2D6) inhibition are commonly used [101,102]. Accordingly, there has been growing interest in developing ML-based tools for the prediction of these criteria. For example, Tian et al. [60] developed a web server called ADMETboost that utilizes the powerful extreme gradient boosting (XGBoost) model to learn about molecule features from multiple fingerprints and descriptors, allowing for the accurate prediction of ADME/T properties, such as Caco2, BBB, CYP2C9 inhibition, CL-Hepa and hERG. It has been demonstrated that this model can achieve remarkable results in the Therapeutics Data Commons ADMET benchmark, ranking first in 18 out of 22 tasks and within the top three in 21 tasks [60]. Similarly, by utilizing more than 13 000 compounds obtained from the PubChem BioAssay Database, Li et al. [65] proposed a multitask autoencoder DNN model to predict the inhibitors of five major cytochrome P450 (CYP450) isoforms (1A2, 2C9, 2C19, 2D6 and 3A4). Especially, the multi-task DNN model achieved average prediction accuracies of 86.4% in 10-fold cross-validation and 88.7% on external test datasets, outperforming single-task models, earlier described classifiers and conventional ML methods [65]. Furthermore, the Tox21 Challenge is a collaborative effort aimed at developing predictive models for toxicity assessment using high-throughput screening data. In this context, Mayr et al. [64] developed a DL pipeline, DeepTox, for toxicity prediction. It outperformed all other computational methods (e.g., naïve Bayes, random forest and support vector machine) in 10 out of 15 cases in the Tox21 Challenge [64]. Therefore, ML algorithms have made significant progress in predicting the ADME/T properties of drugs, contributing to guiding drug safety assessment and preclinical research.

2.3. *Application of ML in Drug Repurposing*

Drug repurposing, also known as drug repositioning, is a strategy to identify new indications from approved or investigational (including failed in clinical trials) drugs that have not been approved [103]. As this approach takes advantage of the extensive safety testing conducted during clinical trials for other purposes, repurposing known drugs not only speeds up the drug development process but also presents cost-saving advantages compared to developing entirely new drugs from scratch [103]. Currently, researchers are increasingly developing and applying ML-based methods to conduct drug repurposing, which can be broadly divided into target-centered and disease-centered approaches [104].

In target-centered drug repurposing, network-based methods have been widely applied to search new targets for known drugs. For example, by employing autoencoder and Positive-Unlabeled matrix completion algorithms, Zeng et al. [70] developed a calculation method called deepDTnet to identify new targets for known drugs from a heterogeneous drug–gene–disease network. Experiments have shown that the deepDTnet achieved a high accuracy in predicting new targets of existing drugs (AUC = 0.963), which is superior to traditional ML methods [70]. Similarly, by combining the network diffusion algorithm and the dimensionality reduction approach, Luo et al. [72] developed DTINet, a novel network-integration procedure for DTI prediction and drug repositioning. It can outperform other existing methods, with AUC and area under precision-recall (AUPR) 5.7% and 5.9% higher than the second best method, respectively, providing an effective tool in the field of drug discovery and target identification [72].

In addition, disease-centered approaches are mainly aimed at identifying drug–disease relationships and can be widely divided into similarity-based and network-based ones [104]. Similarity-based methods have achieved significant progress by combining drug or disease characteristics with the known drug–disease associations [104]. For example, based on the assumption that similar drugs are commonly associated with similar diseases, Luo et al. [73] proposed a novel computational approach called MBiRW, which combines similarity measurements and a Bi-Random walk algorithm to recognize potential novel indications for a specific drug. MBiRW can achieve a high accuracy in predicting drug–disease associations (AUC = 0.917), which is superior to other methods [73]. In addition, network-based methods integrate information from different biological networks to improve the predictive accuracy of drug–disease relationships. For example, Doshi et al. [74] developed a graph neural network model called GDRnet for drug repurposing, which can efficiently screen existing drugs in the database and predict their unknown therapeutic effects by evaluating the scores of drug–disease pairs. Therefore, ML technology holds significant promise in the field of drug repurposing, providing strong support for accelerating drug discovery.

2.4. Application of ML in Chemical Synthesis

Organic synthesis is a key part of the small-molecule drug-discovery process [97]. New molecules are synthesized along the path of compound optimization to achieve improved properties. To promote molecule synthesis, researchers have developed multiple ML-based computational tools applicable to the retrosynthesis prediction and forward reaction prediction.

2.4.1. Retrosynthesis Prediction

Retrosynthesis planning aims to identify efficient synthetic routes for a desired molecule by recursively converting it into easier precursors. Therefore, it can effectively solve the synthesis of complex molecules to facilitate the development of organic synthesis science [105]. At present, a series of ML-based approaches have been used for retrosynthesis planning, mainly including template-based and template-free approaches.

The template-based approach involves systematically comparing the target molecule with a set of templates, each representing alternative substructure patterns that occur during a chemical reaction [105]. The first work involving DNNs for this issue was presented by Segler et al. [79], published in Nature. They found that Monte Carlo tree search (MCTS) combined with DNNs and symbolic rules can be utilized to perform chemical synthesis effectively. The routes generated by the model were comparable to those reported in the literature in a double-blind AB test, thereby confirming the accuracy of the model [79]. However, it is worth noting that template-based approaches cannot be extended beyond templates, limiting their predictive ability [106].

As for the template-free method, it aims to uncover hidden relationships within the data concerning reaction mechanisms rather than relying on direct matching [105]. For example, by using neural sequence-to-sequence models, Liu et al. [80] proposed the template-free method called seq2seq, to perform the retrosynthetic reaction-prediction tasks. This model was based on an encoder–decoder framework consisting of two recurrent neural networks (RNNs) and was trained on a dataset of 50,000 experimental reactions extracted from the United States' patent literature, demonstrating comparable performances to the rule-based expert system model [80]. Therefore, ML algorithms have been extensively employed to conduct retrosynthetic analysis and hold the potential to facilitate chemical synthesis.

2.4.2. Forward Reaction Prediction

Contrary to retrosynthesis analysis, forward reaction prediction aims to identify potential molecules that can be synthesized from given reactants and reagents [105]. Given the reactant molecules as input, the ML model analyzes their structural and chemical properties to generate predictions about the resulting products and reaction conditions. For example, Wei et al. [82] introduced a novel reaction fingerprinting approach that utilizes neural networks to predict reaction types. The prediction results of this method on 16 basic reactions of alkyl halides and alkenes indicates that neural networks can contribute to identify key features from the structure of reactant molecules to classify new reaction types [82]. Similarly, Coley et al. [83] proposed a neural network model to predict the main products of chemical reactions by training the data extracted from a collection of 150,000 compounds' reaction templates in the US patent database. In addition, in practical chemical synthesis reactions, reaction conditions (e.g., solvent and temperature) are critical to maximize the yield of desired products. Based on this, Gao et al. [84] proposed a neural network model to predict the optimal reaction conditions for various types of reactions. This model was trained using a vast dataset of nearly 10 million entries extracted from the Reaxys database and can effectively predict the ideal catalyst, solvent, reagent, and temperature for a given reaction, facilitating the optimization of reaction conditions [84]. Therefore, the utilization of ML-based models can assist in predicting reaction types, accelerating the discovery of new chemical molecules, and identifying optimal reaction conditions, thereby holding great potential in improving the efficiency of chemical synthesis processes.

3. Opportunities for Transformer-Based ML Models in Empowering Drug Discovery

The Transformer model, firstly proposed in the paper 'Attention is All You Need' by Vaswani et al., is a highly advanced DL architecture utilizing self-attention mechanisms. As it allows for parallelization and captures long-range dependencies more efficiently than traditional RNN models, the Transformer model has proven to be highly effective in a wide range of tasks and has set new benchmarks in the corresponding fields [10,11]. Given the advantages of the Transformer, it has emerged as a promising future direction of ML in the field of drug discovery (Figure 3).

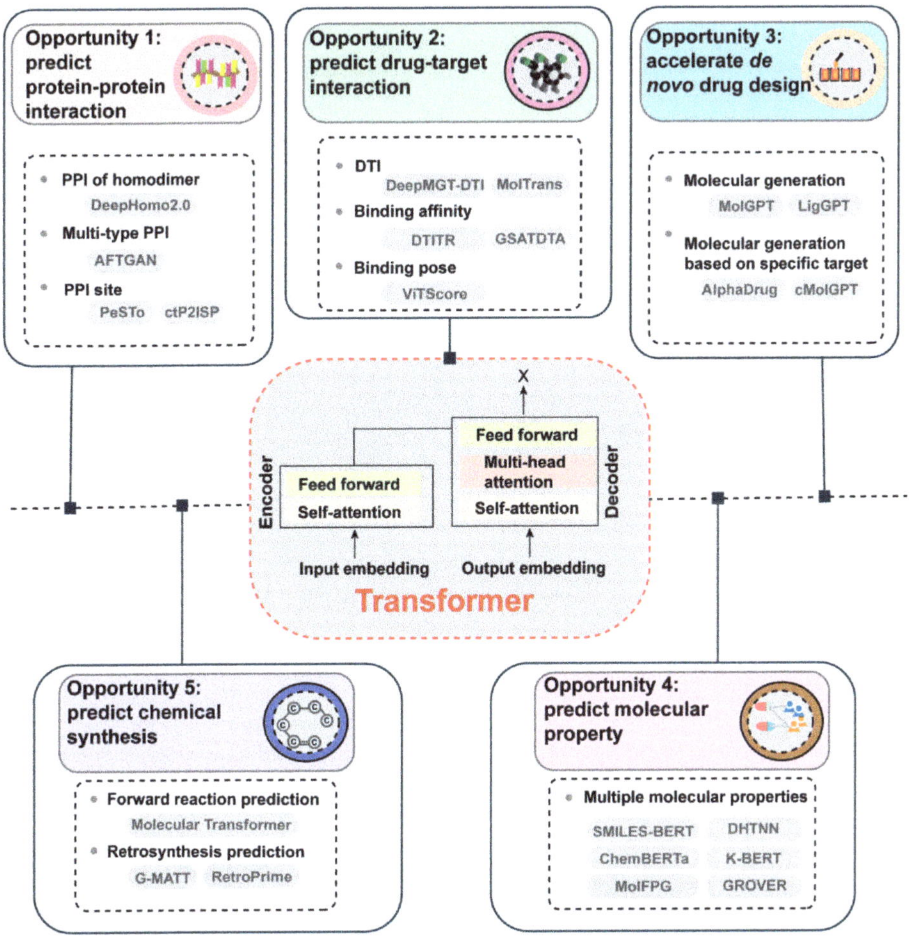

Figure 3. Opportunities for Transformer-based models in empowering drug discovery.

3.1. *Opportunity 1: Transformer Models Empowering PPIs Identification*

Existing ML-based approaches mainly use CNNs to extract low-dimensional features from protein sequences based on the amino acid composition, while disregarding the long-range relationships within these sequences [107]. Fortunately, transformers can capture the long-distance dependencies in the protein sequences, making them suitable to predict whether and how given proteins interact with each other [108]. For example, by utilizing the advantage of the Transformer model in evolutionary scale modeling-multiple sequence alignment, Lin et al. [109] developed DeepHomo2.0, a DL-based model that predicts PPIs of homodimeric complexes by combining Transformer features, monomer structure information, and direct-coupling analysis. The results showed that DeepHomo2.0 can achieve a high accuracy of over 70% and 60% in terms of experimental monomer structure and predicted monomer structure for the top 10 contacts predicted on the Protein Data Bank (PDB)test set, respectively, which is superior to the DCA-based, protein language model-based and other ML-based methods [109]. Similarly, Kang et al. [110] proposed AFTGAN, a neural network that combines Transformer and GAT frameworks for effective protein information extraction and multi-type PPI prediction. Experimental comparisons validated the superior performance of AFTGAN in accurately predicting the PPIs of unknown proteins. Therefore, given the advantage of the Transformer in extracting protein sequences, it has demonstrated remarkable potential in advancing the prediction of PPIs.

3.2. *Opportunity 2: Transformer Models Empowering DTIs' Identification*

Despite the remarkable performance improvement of DL models in DTI prediction, the primary challenge lies in the limited representation of drugs in these methods, as they only consider SMILES sequences, SMARTS strings or molecular graphs, failing to capture comprehensive drug representations [107]. It is worth noting that Transformers can be employed either independently or in combination with other AI algorithms to address these problems. For example, DeepMGT-DTI, a DL model that incorporates a Transformer network and multilayer graph information, can effectively capture the structural features of drugs, leading to improved DTI prediction [111]. Experiments have demonstrated that the DeepMGT-DTI can achieve an AUC of 90.24%, an AUPR of 77.11%, an F1 score of 79.31% and an accuracy of 85.15% on the DrugBank dataset. These performance metrics surpassed those previous target sequence structure models, such as Deep DTA and TransformerCPI [111]. Moreover, GSATDTA, a novel triple-channel model based on graph–sequence attention and Transformer, has been developed to predict the drug-target binding affinity with outstanding performance [107]. Therefore, Transformer models have shown promising results for DTIs' prediction.

3.3. *Opportunity 3: Transformer Models Empowering De Novo Drug Design*

Most existing deep generative models either focus on virtual screening on the available database of compounds by DTI binding-affinity prediction, or unconditionally generate molecules with specific physicochemical and pharmacological properties, which ignore the function of protein targets during the generation process [112]. In contrast, Transformer models have the capability to consider the protein target and achieve target-specific molecular generation. For example, AlphaDrug, a method for protein target-specific *de novo* drug design, has been recently proposed. It utilizes a modified Transformer to optimize the learning of protein information and integrates an efficient MCTS guided by the Transformer's predictions as well as docking values [112]. Notably, in terms of average docking score, uniqueness, the octanol–water partition coefficient logP, the quantitative estimate of drug-likeness (QED), synthetic accessibility (SA) and Natural products-likeness (NP-likeness) criteria, AlphaDrug is superior to other methods (such as LiGANN, SBMolGen and SBDD-3D) [112]. In addition, the GPT model is a powerful language generation model that can be fine-tuned for specific tasks after pre-training on large amounts of text data [113]. It has been successfully applied to accelerate molecular generation for specific targets in the field of drug discovery. For example, cMolGPT, a GPT-inspired model, is a useful tool for target-specific *de novo* molecular generation. The chemical space of the compounds generated by cMolGPT closely matches with that of real target-specific ones [114].

3.4. *Opportunity 4: Transformer Models Empowering Molecular Property Prediction*

Despite the widespread application of ML-based models, the shortage of labeled data continues to be a significant challenge in efficient molecular property predictions [10,115]. To address this, researchers are exploring the use of unlabeled data and leveraging transformer-based self-supervised learning (e.g., BERT) to improve predictions on small-scale labeled data [116]. Currently, several BERT-based pre-training methods for molecular property prediction have been proposed [10,117]. For example, a novel pre-training method, known as K-BERT, was developed to extract chemical information from SMILES similar to chemists for molecular property prediction in drug discovery [118]. The K-BERT model exhibited superior performance in 8 out of 15 tasks, thus reflecting the efficacy and benefits of the proposed pre-training approach in drug discovery. Specifically, K-BERT had an average AUC score of 0.806, outperforming other competing methods (e.g., XGBoost-MACCS, XGBoost-ECFP4, HRGCN+ and Attentive FP) [118]. Moreover, Wang et al. [119] proposed a two-stage (pre-training and fine-tuning) model called SMILES-BERT that could use both unlabeled data and labeled data to improve molecular property prediction. Compared with a range of state-of-the-art approaches (e.g., CircularFP, NeuralFP, Seq2seqFP, Seq3seqFP), it exhibited superior performance on three different datasets (the LogP dataset, PM2 dataset and PCBA-686978 dataset) with accuracies of 0.9154, 0.7589, and 0.8784, respectively [119]. Therefore, these Transformer-based predictors are essential tools for molecular property prediction, contributing to the efficient screening of potential drug candidates.

3.5. *Opportunity 5: Transformer Models Empowering Chemical Synthesis*

Previous sequence-based approaches commonly employed RNNs for both the encoder and decoder, with a single-head attention layer connecting them. These models treated reactants and reagents separately in the input by utilizing atom mapping, which limits the interpretability of the model [120]. Fortunately, Transformer-powered models have shown potential to accelerate chemical synthesis. One notable example is the effectiveness of the multi-head attention Molecular Transformer model in predicting chemical reactions and reaction conditions [120,121]. In addition, inspired by the success of the Molecular Transformer for forward reaction prediction, Schwaller et al. [122] proposed an enhanced Molecular Transformer architecture coupled with a hyper-graph exploration algorithm for automated retrosynthetic pathway prediction. This approach surpasses previous ML-based methods by not only predicting reactants but also identifying reagents for each retrosynthetic step, thereby significantly raising the complexity of the prediction task.

4. Challenges of ML-Based Models in Drug Discovery

Given the remarkable advantages in identifying and extracting features from high-dimensional and complex big data, ML-based models have made significant progress in multiple stages of drug discovery [99]. However, there remain several challenges that have yet to be effectively resolved (Figure 4).

First, the effectiveness of ML algorithms heavily relies on the quantity of training data, and typically, a larger dataset tends to yield a more accurate model [96]. When the amount of data is inadequate, it can significantly impact the performance and reliability of ML models, potentially resulting in the risk of overfitting [123]. Indeed, the limited availability of data, especially labeled data, poses a significant challenge to the progress of ML-driven drug discovery. One potential approach to address this issue is employing transfer learning algorithms, where knowledge acquired from one task can be effectively applied to another task [124–126]. Additionally, in light of the challenges associated with acquiring extensive labeled datasets in drug discovery, there is a growing trend for the effectiveness of concentrating efforts on smaller, carefully curated datasets. This shift highlights the significance of extracting meaningful insights from limited yet relevant data, thereby enhancing the precision and applicability of ML models in the complex landscape of drug discovery.

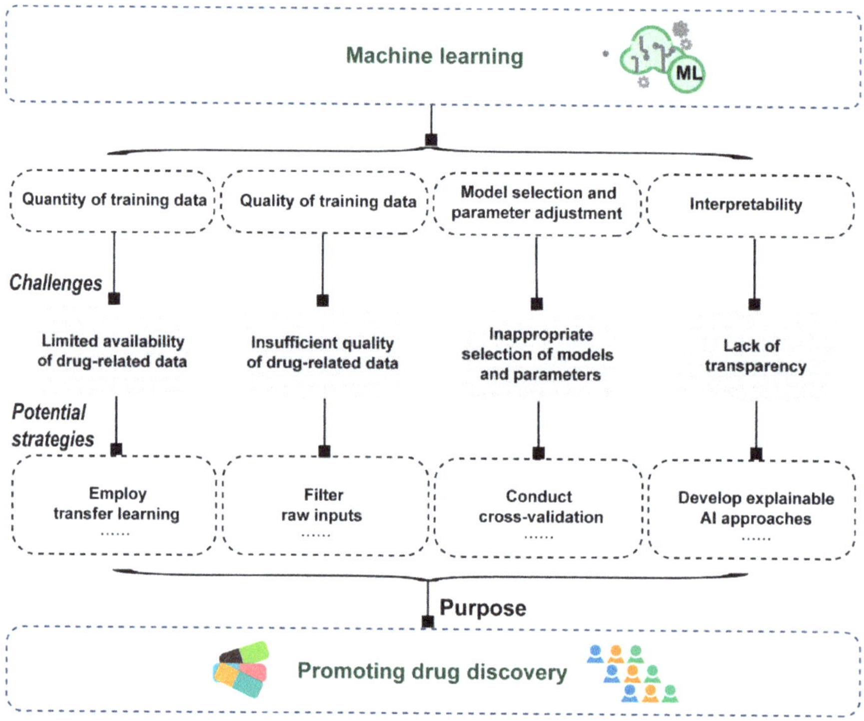

Figure 4. Challenges of machine learning-based models in drug discovery.

Second, the quality of the data is also crucial in determining the prediction performance of ML models. The experimental drug-related data collected in public databases frequently originates from varying biological assays, conditions, or methods, leading to disparate results when different measurement techniques are employed for a specific compound, thereby hindering direct comparisons. Hence, the strategies for filtering raw inputs with noise, outliers, or irrelevant information and automating data entry may be helpful to achieve reliable and accurate ML models for drug discovery. For example, during the data processing phase, noise reduction and outlier detection algorithms, such as Z-scores, box plots or iterative deletion, can be applied to identify and purge outliers from the data, enhancing its quality for ML model prediction. In addition, researchers can use cross-validation experiments to assess the generalization ability of the models, ensuring that they perform well not only on specific datasets but also on new, unseen data.

Third, due to the abundance of ML model architectures and the constant emergence of new ones, it becomes challenging to choose the most suitable models that meet specific research task requirements in the field of drug discovery [99]. Generally, the model selection involves evaluating various options and considering factors such as the complexity of the problem, available data, and computational resources. Furthermore, once the model architecture is selected, the next step is to fine-tune its parameters to optimize the model's performance. Although hyperparameter optimization tools have been proposed to automate the process of tuning substantial parameters in ML models, the entire system process is also relatively complicated, which may bring certain difficulties to the application of researchers [99,127]. In addition, the setting of hyper-parameters usually requires human intervention, which may lead to their incomplete or inaccurate selection. Accordingly, cross-validation is commonly used in variable selection and model parameter tuning to evaluate the performances of various ML methods [128]. Moreover, establishing clear performance metrics at the outset, such as accuracy, precision, recall, F1 score, AUC and AUPR can help in objectively evaluating the suitability of different models depending on the nature of the problem.

Fourth, unlike traditional models where the reasoning and decision-making process can be easily understood, ML models, particularly DL models, operate using complex mathematical algorithms and layers of interconnected neurons, making it challenging to interpret their inner workings. The lack of transparency and interpretability pose difficulties for ML models in explaining the observed phenomena and understanding the underlying biological mechanisms. Hence, the ML models are often referred to as "black boxes" [99]. For this issue, employing visualization tools such as Activation Maximization [129], Local Interpretable Model-agnostic Explanations (LIME) [130] and SHapley Additive exPlanations (SHAP) [131] can help in understanding the model's decision-making process by providing insights into which features are most influential. In the future, a continuous requirement is to develop robust models with high interpretability.

Therefore, a tremendous amount of work has been done to incorporate ML tools to expedite the drug discovery cycle, but further advancement and improvement of these tools is needed before the full potential of ML in drug discovery can be realized.

5. Concluding Remarks

The research and development of new drugs can contribute to meet the human demand for treating diseases and provide more effective, safer, and more convenient treatment options. Compared with the traditional strategies of drug discovery, ML-based approaches have the potential to reduce time and costs, improve safety, and bridge the gap between drug discovery and drug effectiveness, making them increasingly favored by the pharmaceutical industry and academia. In particular, the introduction of chatGPT has sparked researchers' growing interest and exploration in leveraging the Transformer model's NLP capabilities, particularly its self-attention mechanisms, to accelerate multiple stages of the drug discovery process, thereby opening up new opportunities for advancements.

However, the current challenges in ML-based models can result in generating false positives or false negatives, potentially leading to incorrect predictions and resource waste. Further in vitro and in vivo experiments as well as clinical trials are needed to fully demonstrate the practicability of ML-based drug discovery and obtain more reliable and accurate results. Therefore, future research should focus on improving data quality, enhancing the interpretability of ML algorithms, and integrating them with human professional knowledge to increase the efficacy of drug discovery.

Author Contributions: X.Q. and Y.Z. designed the review; Y.Z. collected the related data; X.Q. and Y.Z. drafted the manuscript; X.Q., Y.Z., S.H., Z.Q. and J.C. revised the manuscript. All authors have read and agreed to the published version of the manuscript.

Funding: This work was supported by the National Natural Science Foundation of China (Grant No. 32270705) and the Postgraduate Research & Practice Innovation Program of Jiangsu Province (Grant No. KYCX23_3344).

Institutional Review Board Statement: Not applicable.

Informed Consent Statement: Not applicable.

Data Availability Statement: Not applicable.

Conflicts of Interest: The authors declare no conflicts of interest.

Abbreviations

AI, artificial intelligence; ML, machine learning; DL, deep learning; ANN, artificial neural network; NLP, natural language processing; DTI, drug–target interaction; 3D, three-dimensional; DNN, deep neural network; PPI, protein–protein interaction; GAT, graph attention network; CNN, convolutional neural network; VAE, variational auto-encoder; AAE, adversarial auto-encoder; RNN, recurrent neural network; RF, random forest; XGBoost, eXtreme gradient boosting; GCN, graph convolutional network; SVM, support vector machine; NBC, naïve Bayes classifier; NTN, neural tensor network; GAN, generative adversarial network; GCNN, graph convolutional neural network; ResNet, residual network; GBM, gradient boosting machines; RL, reinforcement learning; GRU, gated recurrent unit; GNN, graph neural networks; k-NN, k-nearest neighbor; LightGBM, light gradient boosting machine; MCTS, Monte Carlo tree search, NTN, neural tensor network; GAN, generative adversarial network; GCNN, graph convolutional neural network; ADME/T, absorption, distribution, metabolism, and excretion and toxicity; BBB, blood–brain barrier; PPB, plasma protein binding, CYP2D6, cytochrome P450 2D6; XGBoost, extreme gradient boosting; CYP450, cytochrome P450;MCTS, Monte Carlo Tree Search; GPT, Generative Pre-Training Transformer; BERT, bidirectional encoder representations from transformers; SC2, Challenge to Predict Aqueous Solubility; MCC, Matthews' Correlation Coefficient; AUC, area under the curve; AUPR, area under precision-recall; PDB, Protein Data Bank; QED, quantitative estimate of drug-likeness; SA, synthetic accessibility;NP, natural products; LIME, Local Interpretable Model-agnostic Explanations; SHAP, SHapley Additive exPlanations.

References

1. DiMasi, J.A.; Grabowski, H.G.; Hansen, R.W. Innovation in the pharmaceutical industry: New estimates of R&D costs. *J. Health Econ.* **2016**, *47*, 20–33. [CrossRef] [PubMed]
2. Ashburn, T.T.; Thor, K.B. Drug repositioning: Identifying and developing new uses for existing drugs. *Nat. Rev. Drug Discov.* **2004**, *3*, 673–683. [CrossRef]
3. Dowden, H.; Munro, J. Trends in clinical success rates and therapeutic focus. *Nat. Rev. Drug Discov.* **2019**, *18*, 495–496. [CrossRef] [PubMed]
4. Deng, J.; Yang, Z.; Ojima, I.; Samaras, D.; Wang, F. Artificial intelligence in drug discovery: Applications and techniques. *Brief. Bioinform* **2022**, *23*, bbab430. [CrossRef] [PubMed]
5. Mak, K.K.; Pichika, M.R. Artificial intelligence in drug development: Present status and future prospects. *Drug Discov. Today* **2019**, *24*, 773–780. [CrossRef]
6. Paul, D.; Sanap, G.; Shenoy, S.; Kalyane, D.; Kalia, K.; Tekade, R.K. Artificial intelligence in drug discovery and development. *Drug Discov. Today* **2021**, *26*, 80–93. [CrossRef]
7. Chen, H.; Engkvist, O.; Wang, Y.; Olivecrona, M.; Blaschke, T. The rise of deep learning in drug discovery. *Drug Discov. Today* **2018**, *23*, 1241–1250. [CrossRef]
8. Wang, K.; Zhou, R.; Li, Y.; Li, M. DeepDTAF: A deep learning method to predict protein-ligand binding affinity. *Brief. Bioinform.* **2021**, *22*, bbab072. [CrossRef]
9. Karimi, M.; Wu, D.; Wang, Z.; Shen, Y. DeepAffinity: Interpretable deep learning of compound-protein affinity through unified recurrent and convolutional neural networks. *Bioinformatics* **2019**, *35*, 3329–3338. [CrossRef]
10. Zhang, S.; Fan, R.; Liu, Y.; Chen, S.; Liu, Q.; Zeng, W. Applications of transformer-based language models in bioinformatics: A survey. *Bioinform. Adv.* **2023**, *3*, vbad001. [CrossRef]
11. Vaswani, A.; Shazeer, N.; Parmar, N.; Uszkoreit, J.; Jones, L.; Gomez, A.N.; Kaiser, Ł.; Polosukhin, I. Attention is all you need. *Adv. Neural Inf. Process. Syst.* **2017**, *30*, 1–11.
12. Kalakoti, Y.; Yadav, S.; Sundar, D. TransDTI: Transformer-Based Language Models for Estimating DTIs and Building a Drug Recommendation Workflow. *ACS Omega* **2022**, *7*, 2706–2717. [CrossRef] [PubMed]
13. Du, Z.; Su, H.; Wang, W.; Ye, L.; Wei, H.; Peng, Z.; Anishchenko, I.; Baker, D.; Yang, J. The trRosetta server for fast and accurate protein structure prediction. *Nat. Protoc.* **2021**, *16*, 5634–5651. [CrossRef]
14. Senior, A.W.; Evans, R.; Jumper, J.; Kirkpatrick, J.; Sifre, L.; Green, T.; Qin, C.; Žídek, A.; Nelson, A.W.R.; Bridgland, A.; et al. Improved protein structure prediction using potentials from deep learning. *Nature* **2020**, *577*, 706–710. [CrossRef]
15. Zhang, L.; Wang, S.; Hou, J.; Si, D.; Zhu, J.; Cao, R. ComplexQA: A deep graph learning approach for protein complex structure assessment. *Brief. Bioinform.* **2023**, *24*, bbad287. [CrossRef] [PubMed]
16. Brandes, N.; Ofer, D.; Peleg, Y.; Rappoport, N.; Linial, M. ProteinBERT: A universal deep-learning model of protein sequence and function. *Bioinformatics* **2022**, *38*, 2102–2110. [CrossRef]
17. Lin, Z.; Akin, H.; Rao, R.; Hie, B.; Zhu, Z.; Lu, W.; Smetanin, N.; Verkuil, R.; Kabeli, O.; Shmueli, Y.; et al. Evolutionary-scale prediction of atomic-level protein structure with a language model. *Science* **2023**, *379*, 1123–1130. [CrossRef]

18. Northey, T.C.; Barešić, A.; Martin, A.C.R. IntPred: A structure-based predictor of protein-protein interaction sites. *Bioinformatics* **2018**, *34*, 223–229. [CrossRef]
19. Maheshwari, S.; Brylinski, M. Template-based identification of protein-protein interfaces using eFindSitePPI. *Methods* **2016**, *93*, 64–71. [CrossRef]
20. Li, Y.; Golding, G.B.; Ilie, L. DELPHI: Accurate deep ensemble model for protein interaction sites prediction. *Bioinformatics* **2021**, *37*, 896–904. [CrossRef]
21. Wang, X.; Zhang, Y.; Yu, B.; Salhi, A.; Chen, R.; Wang, L.; Liu, Z. Prediction of protein-protein interaction sites through eXtreme gradient boosting with kernel principal component analysis. *Comput. Biol. Med.* **2021**, *134*, 104516. [CrossRef]
22. Kang, Y.; Xu, Y.; Wang, X.; Pu, B.; Yang, X.; Rao, Y.; Chen, J. HN-PPISP: A hybrid network based on MLP-Mixer for protein-protein interaction site prediction. *Brief. Bioinform.* **2023**, *24*, bbac480. [CrossRef]
23. Song, B.; Luo, X.; Luo, X.; Liu, Y.; Niu, Z.; Zeng, X. Learning spatial structures of proteins improves protein-protein interaction prediction. *Brief. Bioinform.* **2022**, *23*, bbab558. [CrossRef] [PubMed]
24. Baranwal, M.; Magner, A.; Saldinger, J.; Turali-Emre, E.S.; Elvati, P.; Kozarekar, S.; VanEpps, J.S.; Kotov, N.A.; Violi, A.; Hero, A.O. Struct2Graph: A graph attention network for structure based predictions of protein-protein interactions. *BMC Bioinform.* **2022**, *23*, 370. [CrossRef] [PubMed]
25. Yao, Y.; Du, X.; Diao, Y.; Zhu, H. An integration of deep learning with feature embedding for protein-protein interaction prediction. *PeerJ* **2019**, *7*, e7126. [CrossRef]
26. Huang, Y.; Wuchty, S.; Zhou, Y.; Zhang, Z. SGPPI: Structure-aware prediction of protein-protein interactions in rigorous conditions with graph convolutional network. *Brief. Bioinform.* **2023**, *24*, bbad020. [CrossRef]
27. Du, X.; Sun, S.; Hu, C.; Yao, Y.; Yan, Y.; Zhang, Y. DeepPPI: Boosting Prediction of Protein-Protein Interactions with Deep Neural Networks. *J. Chem. Inf. Model.* **2017**, *57*, 1499–1510. [CrossRef] [PubMed]
28. Wu, J.; Liu, B.; Zhang, J.; Wang, Z.; Li, J. DL-PPI: A method on prediction of sequenced protein-protein interaction based on deep learning. *BMC Bioinform.* **2023**, *24*, 473. [CrossRef]
29. Zhang, F.; Zhang, Y.; Zhu, X.; Chen, X.; Lu, F.; Zhang, X. DeepSG2PPI: A Protein-Protein Interaction Prediction Method Based on Deep Learning. *IEEE/ACM Trans. Comput. Biol. Bioinform.* **2023**, *20*, 2907–2919. [CrossRef] [PubMed]
30. Ghosh, S.; Mitra, P. MaTPIP: A deep-learning architecture with eXplainable AI for sequence-driven, feature mixed protein-protein interaction prediction. *Comput. Methods Programs Biomed.* **2024**, *244*, 107955. [CrossRef]
31. Soleymani, F.; Paquet, E.; Viktor, H.L.; Michalowski, W.; Spinello, D. ProtInteract: A deep learning framework for predicting protein-protein interactions. *Comput. Struct. Biotechnol. J.* **2023**, *21*, 1324–1348. [CrossRef]
32. Cui, Y.; Dong, Q.; Hong, D.; Wang, X. Predicting protein-ligand binding residues with deep convolutional neural networks. *BMC Bioinform.* **2019**, *20*, 93. [CrossRef]
33. Mylonas, S.K.; Axenopoulos, A.; Daras, P. DeepSurf: A surface-based deep learning approach for the prediction of ligand binding sites on proteins. *Bioinformatics* **2021**, *37*, 1681–1690. [CrossRef]
34. Jendele, L.; Krivak, R.; Skoda, P.; Novotny, M.; Hoksza, D. PrankWeb: A web server for ligand binding site prediction and visualization. *Nucleic Acids Res.* **2019**, *47*, W345–w349. [CrossRef]
35. Kandel, J.; Tayara, H.; Chong, K.T. PUResNet: Prediction of protein-ligand binding sites using deep residual neural network. *J. Cheminform.* **2021**, *13*, 65. [CrossRef] [PubMed]
36. Zhou, Y.; Jiang, Y.; Yang, Y. AGAT-PPIS: A novel protein-protein interaction site predictor based on augmented graph attention network with initial residual and identity mapping. *Brief. Bioinform.* **2023**, *24*, bbad122. [CrossRef] [PubMed]
37. Öztürk, H.; Özgür, A.; Ozkirimli, E. DeepDTA: Deep drug-target binding affinity prediction. *Bioinformatics* **2018**, *34*, i821–i829. [CrossRef] [PubMed]
38. He, T.; Heidemeyer, M.; Ban, F.; Cherkasov, A.; Ester, M. SimBoost: A read-across approach for predicting drug-target binding affinities using gradient boosting machines. *J. Cheminform.* **2017**, *9*, 24. [CrossRef]
39. Ahmed, A.; Mam, B.; Sowdhamini, R. DEELIG: A Deep Learning Approach to Predict Protein-Ligand Binding Affinity. *Bioinform. Biol. Insights* **2021**, *15*, 11779322211030364. [CrossRef]
40. Karlov, D.S.; Sosnin, S.; Fedorov, M.V.; Popov, P. graphDelta: MPNN Scoring Function for the Affinity Prediction of Protein-Ligand Complexes. *ACS Omega* **2020**, *5*, 5150–5159. [CrossRef]
41. Feinberg, E.N.; Sur, D.; Wu, Z.; Husic, B.E.; Mai, H.; Li, Y.; Sun, S.; Yang, J.; Ramsundar, B.; Pande, V.S. PotentialNet for Molecular Property Prediction. *ACS Cent. Sci.* **2018**, *4*, 1520–1530. [CrossRef]
42. Liyaqat, T.; Ahmad, T.; Saxena, C. TeM-DTBA: Time-efficient drug target binding affinity prediction using multiple modalities with Lasso feature selection. *J. Comput. Aided Mol. Des.* **2023**, *37*, 573–584. [CrossRef]
43. Wang, C.; Chen, Y.; Zhang, Y.; Li, K.; Lin, M.; Pan, F.; Wu, W.; Zhang, J. A reinforcement learning approach for protein-ligand binding pose prediction. *BMC Bioinform.* **2022**, *23*, 368. [CrossRef]
44. Nguyen, D.D.; Cang, Z.; Wu, K.; Wang, M.; Cao, Y.; Wei, G.W. Mathematical deep learning for pose and binding affinity prediction and ranking in D3R Grand Challenges. *J. Comput. Aided Mol. Des.* **2019**, *33*, 71–82. [CrossRef]
45. Wang, L.; Zhou, Y.; Chen, Q. AMMVF-DTI: A Novel Model Predicting Drug-Target Interactions Based on Attention Mechanism and Multi-View Fusion. *Int. J. Mol. Sci.* **2023**, *24*, 14142. [CrossRef]

46. Popova, M.; Isayev, O.; Tropsha, A. Deep reinforcement learning for de novo drug design. *Sci. Adv.* **2018**, *4*, eaap7885. [CrossRef] [PubMed]
47. Gómez-Bombarelli, R.; Wei, J.N.; Duvenaud, D.; Hernández-Lobato, J.M.; Sánchez-Lengeling, B.; Sheberla, D.; Aguilera-Iparraguirre, J.; Hirzel, T.D.; Adams, R.P.; Aspuru-Guzik, A. Automatic Chemical Design Using a Data-Driven Continuous Representation of Molecules. *ACS Cent. Sci.* **2018**, *4*, 268–276. [CrossRef] [PubMed]
48. Li, Y.; Zhang, L.; Liu, Z. Multi-objective de novo drug design with conditional graph generative model. *J. Cheminform.* **2018**, *10*, 33. [CrossRef] [PubMed]
49. Born, J.; Manica, M.; Oskooei, A.; Cadow, J.; Markert, G.; Rodríguez Martínez, M. PaccMann(RL): De novo generation of hit-like anticancer molecules from transcriptomic data via reinforcement learning. *iScience* **2021**, *24*, 102269. [CrossRef]
50. Kadurin, A.; Nikolenko, S.; Khrabrov, K.; Aliper, A.; Zhavoronkov, A. druGAN: An Advanced Generative Adversarial Autoencoder Model for de Novo Generation of New Molecules with Desired Molecular Properties in Silico. *Mol. Pharm.* **2017**, *14*, 3098–3104. [CrossRef]
51. Coley, C.W.; Rogers, L.; Green, W.H.; Jensen, K.F. SCScore: Synthetic Complexity Learned from a Reaction Corpus. *J. Chem. Inf. Model.* **2018**, *58*, 252–261. [CrossRef] [PubMed]
52. Schoenmaker, L.; Béquignon, O.J.M.; Jespers, W.; van Westen, G.J.P. UnCorrupt SMILES: A novel approach to de novo design. *J. Cheminform.* **2023**, *15*, 22. [CrossRef] [PubMed]
53. Wang, X.; Gao, C.; Han, P.; Li, X.; Chen, W.; Rodríguez Patón, A.; Wang, S.; Zheng, P. PETrans: De Novo Drug Design with Protein-Specific Encoding Based on Transfer Learning. *Int. J. Mol. Sci.* **2023**, *24*, 1146. [CrossRef] [PubMed]
54. Monteiro, N.R.C.; Pereira, T.O.; Machado, A.C.D.; Oliveira, J.L.; Abbasi, M.; Arrais, J.P. FSM-DDTR: End-to-end feedback strategy for multi-objective De Novo drug design using transformers. *Comput. Biol. Med.* **2023**, *164*, 107285. [CrossRef] [PubMed]
55. Song, T.; Ren, Y.; Wang, S.; Han, P.; Wang, L.; Li, X.; Rodriguez-Patón, A. DNMG: Deep molecular generative model by fusion of 3D information for de novo drug design. *Methods* **2023**, *211*, 10–22. [CrossRef]
56. Macedo, B.; Ribeiro Vaz, I.; Taveira Gomes, T. MedGAN: Optimized generative adversarial network with graph convolutional networks for novel molecule design. *Sci. Rep.* **2024**, *14*, 1212. [CrossRef]
57. Panapitiya, G.; Girard, M.; Hollas, A.; Sepulveda, J.; Murugesan, V.; Wang, W.; Saldanha, E. Evaluation of Deep Learning Architectures for Aqueous Solubility Prediction. *ACS Omega* **2022**, *7*, 15695–15710. [CrossRef]
58. Francoeur, P.G.; Koes, D.R. SolTranNet-A Machine Learning Tool for Fast Aqueous Solubility Prediction. *J. Chem. Inf. Model.* **2021**, *61*, 2530–2536. [CrossRef]
59. Zang, Q.; Mansouri, K.; Williams, A.J.; Judson, R.S.; Allen, D.G.; Casey, W.M.; Kleinstreuer, N.C. In Silico Prediction of Physicochemical Properties of Environmental Chemicals Using Molecular Fingerprints and Machine Learning. *J. Chem. Inf. Model.* **2017**, *57*, 36–49. [CrossRef]
60. Tian, H.; Ketkar, R.; Tao, P. ADMETboost: A web server for accurate ADMET prediction. *J. Mol. Model.* **2022**, *28*, 408. [CrossRef]
61. Schyman, P.; Liu, R.; Desai, V.; Wallqvist, A. vNN Web Server for ADMET Predictions. *Front. Pharmacol.* **2017**, *8*, 889. [CrossRef] [PubMed]
62. Wei, Y.; Li, S.; Li, Z.; Wan, Z.; Lin, J. Interpretable-ADMET: A web service for ADMET prediction and optimization based on deep neural representation. *Bioinformatics* **2022**, *38*, 2863–2871. [CrossRef] [PubMed]
63. Deng, D.; Chen, X.; Zhang, R.; Lei, Z.; Wang, X.; Zhou, F. XGraphBoost: Extracting Graph Neural Network-Based Features for a Better Prediction of Molecular Properties. *J. Chem. Inf. Model.* **2021**, *61*, 2697–2705. [CrossRef] [PubMed]
64. Mayr, A.; Klambauer, G.; Unterthiner, T.; Hochreiter, S. DeepTox: Toxicity prediction using deep learning. *Front. Environ. Sci.* **2016**, *3*, 80. [CrossRef]
65. Li, X.; Xu, Y.; Lai, L.; Pei, J. Prediction of Human Cytochrome P450 Inhibition Using a Multitask Deep Autoencoder Neural Network. *Mol. Pharm.* **2018**, *15*, 4336–4345. [CrossRef]
66. Shaker, B.; Yu, M.S.; Song, J.S.; Ahn, S.; Ryu, J.Y.; Oh, K.S.; Na, D. LightBBB: Computational prediction model of blood-brain-barrier penetration based on LightGBM. *Bioinformatics* **2021**, *37*, 1135–1139. [CrossRef]
67. Tang, Q.; Nie, F.; Zhao, Q.; Chen, W. A merged molecular representation deep learning method for blood-brain barrier permeability prediction. *Brief. Bioinform.* **2022**, *23*, bbac357. [CrossRef]
68. Jang, W.D.; Jang, J.; Song, J.S.; Ahn, S.; Oh, K.S. PredPS: Attention-based graph neural network for predicting stability of compounds in human plasma. *Comput. Struct. Biotechnol. J.* **2023**, *21*, 3532–3539. [CrossRef]
69. Khaouane, A.; Khaouane, L.; Ferhat, S.; Hanini, S. Deep Learning for Drug Development: Using CNNs in MIA-QSAR to Predict Plasma Protein Binding of Drugs. *AAPS PharmSciTech* **2023**, *24*, 232. [CrossRef] [PubMed]
70. Zeng, X.; Zhu, S.; Lu, W.; Liu, Z.; Huang, J.; Zhou, Y.; Fang, J.; Huang, Y.; Guo, H.; Li, L.; et al. Target identification among known drugs by deep learning from heterogeneous networks. *Chem. Sci.* **2020**, *11*, 1775–1797. [CrossRef] [PubMed]
71. Wan, F.; Hong, L.; Xiao, A.; Jiang, T.; Zeng, J. NeoDTI: Neural integration of neighbor information from a heterogeneous network for discovering new drug-target interactions. *Bioinformatics* **2019**, *35*, 104–111. [CrossRef]
72. Luo, Y.; Zhao, X.; Zhou, J.; Yang, J.; Zhang, Y.; Kuang, W.; Peng, J.; Chen, L.; Zeng, J. A network integration approach for drug-target interaction prediction and computational drug repositioning from heterogeneous information. *Nat. Commun.* **2017**, *8*, 573. [CrossRef]

73. Luo, H.; Wang, J.; Li, M.; Luo, J.; Peng, X.; Wu, F.X.; Pan, Y. Drug repositioning based on comprehensive similarity measures and Bi-Random walk algorithm. *Bioinformatics* **2016**, *32*, 2664–2671. [CrossRef]
74. Doshi, S.; Chepuri, S.P. A computational approach to drug repurposing using graph neural networks. *Comput. Biol. Med.* **2022**, *150*, 105992. [CrossRef] [PubMed]
75. Zeng, X.; Zhu, S.; Liu, X.; Zhou, Y.; Nussinov, R.; Cheng, F. deepDR: A network-based deep learning approach to in silico drug repositioning. *Bioinformatics* **2019**, *35*, 5191–5198. [CrossRef]
76. Jiang, H.J.; Huang, Y.A.; You, Z.H. Predicting Drug-Disease Associations via Using Gaussian Interaction Profile and Kernel-Based Autoencoder. *BioMed Res. Int.* **2019**, *2019*, 2426958. [CrossRef]
77. Ghorbanali, Z.; Zare-Mirakabad, F.; Salehi, N.; Akbari, M.; Masoudi-Nejad, A. DrugRep-HeSiaGraph: When heterogenous siamese neural network meets knowledge graphs for drug repurposing. *BMC Bioinform.* **2023**, *24*, 374. [CrossRef] [PubMed]
78. Suviriyapaisal, N.; Wichadakul, D. iEdgeDTA: Integrated edge information and 1D graph convolutional neural networks for binding affinity prediction. *RSC Adv.* **2023**, *13*, 25218–25228. [CrossRef]
79. Segler, M.H.S.; Preuss, M.; Waller, M.P. Planning chemical syntheses with deep neural networks and symbolic AI. *Nature* **2018**, *555*, 604–610. [CrossRef]
80. Liu, B.; Ramsundar, B.; Kawthekar, P.; Shi, J.; Gomes, J.; Luu Nguyen, Q.; Ho, S.; Sloane, J.; Wender, P.; Pande, V. Retrosynthetic Reaction Prediction Using Neural Sequence-to-Sequence Models. *ACS Cent. Sci.* **2017**, *3*, 1103–1113. [CrossRef]
81. Thakkar, A.; Chadimová, V.; Bjerrum, E.J.; Engkvist, O.; Reymond, J.L. Retrosynthetic accessibility score (RAscore)—Rapid machine learned synthesizability classification from AI driven retrosynthetic planning. *Chem. Sci.* **2021**, *12*, 3339–3349. [CrossRef]
82. Wei, J.N.; Duvenaud, D.; Aspuru-Guzik, A. Neural Networks for the Prediction of Organic Chemistry Reactions. *ACS Cent. Sci.* **2016**, *2*, 725–732. [CrossRef]
83. Coley, C.W.; Barzilay, R.; Jaakkola, T.S.; Green, W.H.; Jensen, K.F. Prediction of Organic Reaction Outcomes Using Machine Learning. *ACS Cent. Sci.* **2017**, *3*, 434–443. [CrossRef]
84. Gao, H.; Struble, T.J.; Coley, C.W.; Wang, Y.; Green, W.H.; Jensen, K.F. Using Machine Learning To Predict Suitable Conditions for Organic Reactions. *ACS Cent. Sci.* **2018**, *4*, 1465–1476. [CrossRef]
85. Marcou, G.; Aires de Sousa, J.; Latino, D.A.; de Luca, A.; Horvath, D.; Rietsch, V.; Varnek, A. Expert system for predicting reaction conditions: The Michael reaction case. *J. Chem. Inf. Model.* **2015**, *55*, 239–250. [CrossRef]
86. You, Z.H.; Li, S.; Gao, X.; Luo, X.; Ji, Z. Large-scale protein-protein interactions detection by integrating big biosensing data with computational model. *BioMed Res. Int.* **2014**, *2014*, 598129. [CrossRef]
87. Chan, H.C.S.; Shan, H.; Dahoun, T.; Vogel, H.; Yuan, S. Advancing Drug Discovery via Artificial Intelligence. *Trends Pharmacol. Sci.* **2019**, *40*, 592–604. [CrossRef]
88. Muhammed, M.T.; Aki-Yalcin, E. Homology modeling in drug discovery: Overview, current applications, and future perspectives. *Chem. Biol. Drug Des.* **2019**, *93*, 12–20. [CrossRef]
89. Zhang, Y.; Skolnick, J. The protein structure prediction problem could be solved using the current PDB library. *Proc. Natl. Acad. Sci. USA* **2005**, *102*, 1029–1034. [CrossRef]
90. Tang, T.; Zhang, X.; Liu, Y.; Peng, H.; Zheng, B.; Yin, Y.; Zeng, X. Machine learning on protein-protein interaction prediction: Models, challenges and trends. *Brief. Bioinform.* **2023**, *24*, bbad076. [CrossRef]
91. Soleymani, F.; Paquet, E.; Viktor, H.; Michalowski, W.; Spinello, D. Protein-protein interaction prediction with deep learning: A comprehensive review. *Comput. Struct. Biotechnol. J.* **2022**, *20*, 5316–5341. [CrossRef]
92. Li, S.; Wu, S.; Wang, L.; Li, F.; Jiang, H.; Bai, F. Recent advances in predicting protein-protein interactions with the aid of artificial intelligence algorithms. *Curr. Opin. Struct. Biol.* **2022**, *73*, 102344. [CrossRef]
93. Tripathi, N.; Goshisht, M.K.; Sahu, S.K.; Arora, C. Applications of artificial intelligence to drug design and discovery in the big data era: A comprehensive review. *Mol. Divers.* **2021**, *25*, 1643–1664. [CrossRef]
94. Dhakal, A.; McKay, C.; Tanner, J.J.; Cheng, J. Artificial intelligence in the prediction of protein-ligand interactions: Recent advances and future directions. *Brief. Bioinform.* **2022**, *23*, bbab476. [CrossRef]
95. Nicolaou, C.A.; Kannas, C.; Loizidou, E. Multi-objective optimization methods in de novo drug design. *Mini Rev. Med. Chem.* **2012**, *12*, 979–987. [CrossRef]
96. Zhong, F.; Xing, J.; Li, X.; Liu, X.; Fu, Z.; Xiong, Z.; Lu, D.; Wu, X.; Zhao, J.; Tan, X.; et al. Artificial intelligence in drug design. *Sci. China. Life Sci.* **2018**, *61*, 1191–1204. [CrossRef]
97. Hessler, G.; Baringhaus, K.H. Artificial Intelligence in Drug Design. *Molecules* **2018**, *23*, 2520. [CrossRef]
98. Schneider, G.; Funatsu, K.; Okuno, Y.; Winkler, D. De novo Drug Design—Ye olde Scoring Problem Revisited. *Mol. Inform.* **2017**, *36*, 1681031. [CrossRef]
99. Wang, L.; Ding, J.; Pan, L.; Cao, D.; Jiang, H.; Ding, X. Artificial intelligence facilitates drug design in the big data era. *Chemom. Intell. Lab. Syst.* **2019**, *194*, 103850. [CrossRef]
100. Yang, X.; Wang, Y.; Byrne, R.; Schneider, G.; Yang, S. Concepts of Artificial Intelligence for Computer-Assisted Drug Discovery. *Chem. Rev.* **2019**, *119*, 10520–10594. [CrossRef]
101. Yu, E.; Xu, Y.; Shi, Y.; Yu, Q.; Liu, J.; Xu, L. Discovery of novel natural compound inhibitors targeting estrogen receptor α by an integrated virtual screening strategy. *J. Mol. Model.* **2019**, *25*, 278. [CrossRef]

102. Zhong, W.; Zhao, L.; Yang, Z.; Yu-Chian Chen, C. Graph convolutional network approach to investigate potential selective Limk1 inhibitors. *J. Mol. Graph. Model.* **2021**, *107*, 107965. [CrossRef]
103. Zhou, Y.; Wang, F.; Tang, J.; Nussinov, R.; Cheng, F. Artificial intelligence in COVID-19 drug repurposing. *Lancet Digit. Health* **2020**, *2*, e667–e676. [CrossRef]
104. Pan, X.; Lin, X.; Cao, D.; Zeng, X.; Yu, P.S.; He, L.; Nussinov, R.; Cheng, F. Deep learning for drug repurposing: Methods, databases, and applications. *Wiley Interdiscip. Rev. Comput. Mol. Sci.* **2022**, *12*, e1597. [CrossRef]
105. Dong, J.; Zhao, M.; Liu, Y.; Su, Y.; Zeng, X. Deep learning in retrosynthesis planning: Datasets, models and tools. *Brief. Bioinform.* **2022**, *23*, bbab391. [CrossRef] [PubMed]
106. Lee, A.A.; Yang, Q.; Sresht, V.; Bolgar, P.; Hou, X.; Klug-McLeod, J.L.; Butler, C.R. Molecular Transformer unifies reaction prediction and retrosynthesis across pharma chemical space. *Chem. Commun.* **2019**, *55*, 12152–12155. [CrossRef]
107. Yan, X.; Liu, Y. Graph-sequence attention and transformer for predicting drug-target affinity. *RSC Adv.* **2022**, *12*, 29525–29534. [CrossRef]
108. Lee, M. Recent Advances in Deep Learning for Protein-Protein Interaction Analysis: A Comprehensive Review. *Molecules* **2023**, *28*, 5169. [CrossRef]
109. Lin, P.; Yan, Y.; Huang, S.Y. DeepHomo2.0: Improved protein-protein contact prediction of homodimers by transformer-enhanced deep learning. *Brief. Bioinform.* **2023**, *24*, bbac499. [CrossRef] [PubMed]
110. Kang, Y.; Elofsson, A.; Jiang, Y.; Huang, W.; Yu, M.; Li, Z. AFTGAN: Prediction of multi-type PPI based on attention free transformer and graph attention network. *Bioinformatics* **2023**, *39*, btad052. [CrossRef]
111. Zhang, P.; Wei, Z.; Che, C.; Jin, B. DeepMGT-DTI: Transformer network incorporating multilayer graph information for Drug-Target interaction prediction. *Comput. Biol. Med.* **2022**, *142*, 105214. [CrossRef]
112. Qian, H.; Lin, C.; Zhao, D.; Tu, S.; Xu, L. AlphaDrug: Protein target specific de novo molecular generation. *PNAS Nexus* **2022**, *1*, pgac227. [CrossRef]
113. Radford, A.; Narasimhan, K.; Salimans, T.; Sutskever, I. Improving Language Understanding by Generative Pre-Training. 2018. Available online: https://www.mikecaptain.com/resources/pdf/GPT-1.pdf (accessed on 14 January 2024).
114. Wang, Y.; Zhao, H.; Sciabola, S.; Wang, W. cMolGPT: A Conditional Generative Pre-Trained Transformer for Target-Specific De Novo Molecular Generation. *Molecules* **2023**, *28*, 4430. [CrossRef]
115. Chithrananda, S.; Grand, G.; Ramsundar, B. ChemBERTa: Large-scale self-supervised pretraining for molecular property prediction. *arXiv* **2020**, arXiv:2010.09885.
116. Devlin, J.; Chang, M.-W.; Lee, K.; Toutanova, K. Bert: Pre-training of deep bidirectional transformers for language understanding. *arXiv* **2018**, arXiv:1810.04805.
117. Liu, Z.; Roberts, R.A.; Lal-Nag, M.; Chen, X.; Huang, R.; Tong, W. AI-based language models powering drug discovery and development. *Drug Discov. Today* **2021**, *26*, 2593–2607. [CrossRef] [PubMed]
118. Wu, Z.; Jiang, D.; Wang, J.; Zhang, X.; Du, H.; Pan, L.; Hsieh, C.Y.; Cao, D.; Hou, T. Knowledge-based BERT: A method to extract molecular features like computational chemists. *Brief. Bioinform.* **2022**, *23*, bbac131. [CrossRef]
119. Wang, S.; Guo, Y.; Wang, Y.; Sun, H.; Huang, J. Smiles-bert: Large scale unsupervised pre-training for molecular property prediction. In Proceedings of the 10th ACM International Conference on Bioinformatics, Computational Biology and Health Informatics, Niagara Falls, NY, USA, 7–10 September 2019; pp. 429–436.
120. Schwaller, P.; Laino, T.; Gaudin, T.; Bolgar, P.; Hunter, C.A.; Bekas, C.; Lee, A.A. Molecular Transformer: A Model for Uncertainty-Calibrated Chemical Reaction Prediction. *ACS Cent. Sci.* **2019**, *5*, 1572–1583. [CrossRef]
121. Andronov, M.; Voinarovska, V.; Andronova, N.; Wand, M.; Clevert, D.-A.; Schmidhuber, J. Reagent prediction with a molecular transformer improves reaction data quality. *Chem. Sci.* **2023**, *14*, 3235–3246. [CrossRef]
122. Schwaller, P.; Petraglia, R.; Zullo, V.; Nair, V.H.; Haeuselmann, R.A.; Pisoni, R.; Bekas, C.; Iuliano, A.; Laino, T. Predicting retrosynthetic pathways using transformer-based models and a hyper-graph exploration strategy. *Chem. Sci.* **2020**, *11*, 3316–3325. [CrossRef]
123. Mamoshina, P.; Vieira, A.; Putin, E.; Zhavoronkov, A. Applications of Deep Learning in Biomedicine. *Mol. Pharm.* **2016**, *13*, 1445–1454. [CrossRef]
124. Lu, J.; Wang, C.; Zhang, Y. Predicting Molecular Energy Using Force-Field Optimized Geometries and Atomic Vector Representations Learned from an Improved Deep Tensor Neural Network. *J. Chem. Theory Comput.* **2019**, *15*, 4113–4121. [CrossRef]
125. Cai, C.; Wang, S.; Xu, Y.; Zhang, W.; Tang, K.; Ouyang, Q.; Lai, L.; Pei, J. Transfer Learning for Drug Discovery. *J. Med. Chem.* **2020**, *63*, 8683–8694. [CrossRef]
126. Altae-Tran, H.; Ramsundar, B.; Pappu, A.S.; Pande, V. Low Data Drug Discovery with One-Shot Learning. *ACS Cent. Sci.* **2017**, *3*, 283–293. [CrossRef]
127. Krizhevsky, A.; Sutskever, I.; Hinton, G.E. ImageNet classification with deep convolutional neural networks. *Commun. ACM* **2017**, *60*, 84–90. [CrossRef]
128. Chang, M. AI for Drug Development and Well-Being. 2020. Available online: http://ctrisoft.net/StatisticiansOrg/AI/AIforWellbingebook5.5x8.5in.pdf (accessed on 14 January 2024).
129. Erhan, D.; Bengio, Y.; Courville, A.C.; Vincent, P. *Visualizing Higher-Layer Features of a Deep Network*; University of Montreal: Montreal, QC, USA, 2009.

130. Ribeiro, M.T.; Singh, S.; Guestrin, C. "Why Should I Trust You?": Explaining the Predictions of Any Classifier. In Proceedings of the 22nd ACM SIGKDD International Conference on Knowledge Discovery and Data Mining, San Francisco, CA, USA, 13–17 August 2016; pp. 1135–1144.
131. Lundberg, S.M.; Lee, S.-I. A Unified Approach to Interpreting Model Predictions. In Proceedings of the Neural Information Processing Systems, Long Beach, CA, USA, 4–9 December 2017.

MDPI AG
Grosspeteranlage 5
4052 Basel
Switzerland
Tel.: +41 61 683 77 34

Molecules Editorial Office
E-mail: molecules@mdpi.com
www.mdpi.com/journal/molecules

www.ingramcontent.com/pod-product-compliance
Lightning Source LLC
Chambersburg PA
CBHW062052220926
51850CB00010B/2743
* 9 7 8 3 7 2 5 8 6 7 0 2 8 *